PROGRAMME MANAGEMENT
FOR OWNER TEAMS

This page intentionally left blank

PROGRAMME MANAGEMENT FOR OWNER TEAMS

A PRACTICAL GUIDE TO WHAT YOU NEED TO KNOW

FREEK VAN HEERDEN
JURIE STEYN &
DAVIDA VAN DER WALT

JULY 2015
AN OTC PUBLICATION

Programme Management for Owner Teams – a practical guide to what you need to know

For information contact:
Owner Team Consultation (Pty) Ltd
PO Box 60601, Vaalpark, 1948, South Africa

www.ownerteamconsult.com

Registered versions of this publication:

ISBN: 978-0-620-65837-9 (Paperback) – this version

ISBN: 978-0-620-65838-6 (Pdf)

ISBN: 978-0-620-65839-3 (ePub)

Cover photo credits:
Construction site – 123RF.com stock photo
Chemical facility – Dollar Photo Club stock photo
Squacco heron – Deryck Coetzer (www.deryckphotoworld.com)

Table of Contents

-----/////-----

vii

Foreword

Successful large industrial and infrastructure programmes and the projects they comprise can be key contributors to the growth of companies, as well as the national and regional economies they support. They can be significant enablers of growth in other industries and the economy in general. They are also, arguably, more the exception than the norm.

In this timely and important contribution, Freek van Heerden, Jurie Steyn and Davida van der Walt provide a practical guide to programme success for the owner team - the people who will conceive, shape, plan and manage the execution of the programme and will continuously be held accountable over long periods of time, for meeting its objectives. They are the team that will develop the initial concept long before other stakeholders are engaged, and will account ultimately to stakeholders in meeting programme objectives and demonstrating strategic benefits achieved, or not, as the case may be.

I have worked with the authors on many projects over my entire career and I can vouch for their grasp of project management and the importance of strong owner teams in shaping and executing successful programmes, as well as their understanding of the enabling factors – those critical to ensuring these teams are kept engaged with their leadership, each other and their stakeholders over extensive periods of time, and all this in a changing business environment. We could each of us relate examples of projects we have worked on where the principles and guidance provided in this book either made for a successful outcome, or may have, if only they had been followed.

I can certainly endorse the notion that strong, well organised owner teams, following a well-defined programme strategy, are vitally necessary, if not sufficient, for successful programme outcomes. Of course programme management and engineering and construction contractors like the one I have spent the past 20 years with, and others in the project supply chains are also critical to project and programme success. We are likewise necessary, but not sufficient for success without a well organised owner team.

The authors have put forward a very useful management model customised to the needs of a complex programme with a long time span. This covers the programme lifecycle in multiple phases from the initial strategic need and idea generation phase, through planning, execution to the ultimate and critical review of whether the programme has met its objectives and strategic goals in the face of a lifespan of years and continual changes in the business environment. I found the use of a very plausible and detailed case study extremely useful in giving life to the concepts introduced by the model, and the associated case sample documents such as a programme charter and mandate complete the practical application of the concepts.

Importantly, the people and communication factors that enable the long-term, phased programme execution introduced in the first parts of the book are not ignored. Too often management teams will focus first and foremost on the technical and business tracks of a new venture, and less on the lifecycle enablers and the well-being of the team and on communication and engagement with other stakeholders. When not built into programme execution plans, the neglect of these factors could be disastrous to the programme and the team, or individual team members. The model provides a very clear demonstration of the importance of the leadership, communication, engagement and governance aspects of a programme business plan with practical guidance on each.

While the book is aimed primarily at owner teams charged with initiating and delivering complex programmes, it has relevance to other stakeholders, both internal in the owner organisation as well as external. This is a must-read, not only for project organisations, but also for their operating company counterparts, boards of directors and their managing and other contractors, and other service providers alike. I find relevance of the principles outlined also to the operation of a project delivery company, even though in that case the projects are unrelated to each other from an ownership perspective, and there are multiple owner teams involved as our clients.

It seems there are always examples in industry and nationally of programmes facing major challenges and irrecoverable delays and over-runs. These very large developments consume large amounts of capital and other resources. If industry statistics are believed, and megaprojects are more likely to fail than to meet their objectives, then this contribution to programme management is an essential starting point, and its management framework and practical, case-based guidance is bound to assist owner teams in delivering more successful outcomes. This, in turn, must make a contribution to the effective allocation of resources, to economic growth and to job creation so sorely needed in many economies.

Mark Flower
Vice President, Fluor South Africa
Johannesburg

April, 2015

This page intentionally left blank

Typical Programme Challenges

Managing the shaping process to fix the programme scope

On a large programme, which extends over a long time period, the tendency is often to keep on adding new business objectives to the programme scope, resulting in ever moving targets, costs and timelines. The first few chapters in the book covers programme shaping and planning best practices. Recommendations are made on how to close out the programme scope timeously and how to prevent scope creep thereafter.

Structuring an effective owner programme management team

An owner organisation wishing to implement a programme is well advised to mobilise a strong owner team, consisting of the owner's own personnel. It is often said that the owner does not have sufficient competent personnel to staff an owner team for a large programme. The fall-back position is then to appoint a managing contractor to manage the programme on behalf of the owner. It has been shown that this approach very rarely results in a successful programme. Guidance is provided in chapter 6 on how to develop an effective owner team whilst ensuring that the owner organisation retains overall control.

Ensuring focus on business objectives

It is often difficult to demonstrate that a programme has met its objectives at the end of the programme lifecycle. During the life of a programme, the affected business units are normally implementing various other initiatives that might affect the baseline performance. The original baseline is used at the end of the programme to measure the impact thereof, but because this baseline has moved, the outcome is normally unclear. Effective baseline tracking, supported by rigorous benefits management, as described in

Chapters 8 and 9 may well indicate that the programme met its objectives.

Keeping internal stakeholders aligned and informed

Internal stakeholders sometimes become disillusioned and start interfering directly in the management of a programme. Lack of transparency causes unease and therefore the need to intervene (often unnecessarily or too late). Developing an effective monitoring system that tracks progress against the programme and business objectives, coupled with a detailed stakeholder management and communication plan (Chapter 12), helps alleviate the concerns of stakeholders. It also enables them to provide pro-active input and guidance on an informed basis.

Keeping the programme team motivated and engaged

Employees are assigned to a programme for an extended duration and then 'forgotten' in terms of personal development. In addition, a programme is a stressful environment, leading to exhaustion and burn-out. These problems can be overcome by effective leadership combined with good management practices. This coupled with team health monitoring and support, goes a long way towards keeping the team motivated. Nurturing (Chapter 11) and alignment (Chapter 13) are key elements in our programme management model.

Ensuring effective programme governance

A company cannot be sustainable without effective governance. Similarly, a programme must also be subject to governance under the accountability of the programme sponsor. However, governance can be overcomplicated, leading to bureaucracy and long decision making chains. Chapter 14 addresses governance requirements in general, linking these to programme governance requirements and the roles and accountabilities of the various bodies within a programme environment.

Preface

Introduction

The intent of this book is to summarise what programme management entails. It provides a practical overview and understanding of what programme management is compared to managing a single project or a portfolio of projects. This is done from an owner team perspective.

The book provides guidelines, case study examples, as well as the tools required in all the relevant areas of project and programme management. Programmes are covered from initial inception through to final completion, focusing not only on achieving project objectives, but also on the overall business objectives of the programme. The focus and level of detail required from an owner programme management team are covered.

The book covers programme initiation and shaping, business ethics, obtaining alignment amongst stakeholders, planning, organising and control aspects, communication and nurturing of the team throughout the lifecycle of the programme.

For whom is the book intended?

This is not a book on project management. The premise is that readers are familiar with the terminology and practice of project management and are seeking to advance their understanding of programme management.

An important factor is that the book is written from the viewpoint of the owner organisation and not from the viewpoint of engineering and project management contractors. The owner organisation is that entity for whom the project or programme is being implemented and who will own, operate and manage it in future. Hence the title: *Programme management for owner teams – A practical guide to what you need to know.*

What background knowledge is needed?

Readers should have a sound understanding of project management and should ideally be in a position where they manage, or are responsible for, multiple projects in an owner organisation. It is not essential to have background knowledge of engineering or strategic business management, although such knowledge will be beneficial.

Companion websites

Our websites are sources of articles, templates and procedures that can be accessed to simplify the task of programme management.

The first is our company website which is open to all and is available at www.ownerteamconsult.com. Apart from the normal company and marketing information, this site is used as the vehicle of our monthly Insight Articles, covering a range of project, programme and business related topics.

The second website is available at www.otctoolkits.com and is the repository of our project and programme management templates, procedures and training modules. Selected content is available to all registered owners of this book. We intend to continually add to the content on the websites and update the contents when appropriate.

Layout of the book

The book comprises 15 chapters, divided into three parts.

Part 1 sets the scene and contains three chapters. Chapter 1 is an overview of essential concepts regarding project and programme management to ensure that a common and consistent terminology is established. Chapter 2 describes the owner organisation and what is required from the owner during a programme. A model for programme management is presented in Chapter 3. A case study is also introduced in Chapter 3, which is used and built on throughout the book.

Part 2 focusses on the six sequential programme steps as defined in the programme management model. Each of the steps has a chapter devoted to it and the application of the steps is illustrated using the case study from Chapter 3. The principle is to have a good split between theory and practical examples to ensure that the concepts presented are well understood and entrenched.

Part 3 deals with what we refer to as 'programme lifecycle essentials' or matters that require attention throughout the lifecycle of the programme. This includes aspects such as leadership, nurturing, stakeholder engagement, communications, alignment and programme governance. Finally, we attempt to answer whether a programme will be a guaranteed success if the principles presented in this book are followed.

About the authors

The authors have been involved in the petrochemical industry for an extensive period of time, mainly in developing and implementing projects in various parts of the world. Collective experience also includes business management, operations management, engineering management and social dynamics. They have been involved as project team members of the owner organisation in small, large and megaprojects, project portfolios and project programmes.

Freek van Heerden

Frederick Jacobus van Heerden has been involved in the petrochemical industry since graduating as a chemical engineer in 1976 and is a seasoned project professional on megaprojects and programmes. He has worked on projects ranging from wholly-owned projects, to joint ventures, mergers and acquisitions. In recent years, he filled positions as senior technical director on projects in an owner team capacity, as well as in business development, general management and engineering management.

He has also been involved in chemical engineering education for many years. He has vast experience of scoping and alignment on projects and in value assurance. Freek is very interested in understanding and developing owner organisation project team capabilities such that they can better fulfil the role of an owner project team and not just replicate the roles and responsibilities of the engineering contractor.

Jurie Steyn

Jurie Wynand Steyn holds degrees in chemical engineering and an MBA. He has extensive experience as gleaned from a working career of over 35 years covering process development, plant process engineering, project management, operations management and environmental, health and safety engineering and management. Operations management was primarily in Fischer–Tropsch facilities, water and effluent treatment, Phenosolvan and phenol and cresol plants. He was also responsible for production planning, coordination and information systems for several years.

He has been involved internationally in environmental remediation activities and environmental due diligence evaluations. He was directly involved in environmental, health and safety management during the construction phase of projects, land risk management and remediation of land contaminated by chemical spills and emissions. Jurie has a healthy interest in business ethics and believes that an ethical business is a sustainable business.

Davida van der Walt

Davida van der Walt, a registered industrial psychologist, is a project professional who brings a non-technical view to projects. She has vast experience in facilitating alignment processes on projects such as business and technology strategy facilitation, framing and alignment processes, roles and responsibilities workshops, communication strategies, problem solving and lessons learnt. Her exposure on projects covers social responsibility projects, enterprise development projects, megaprojects and programmes.

Her understanding of people dynamics in dealing with project stakeholders adds significant value in ensuring smooth project planning and execution. Specific areas of expertise include project and programme team wellness, team alignment and stakeholder communication.

Symbology

Throughout the book we make use of a combination of symbols and different text colours to highlight aspects of the text. The following three sets of symbols and text colours are used:

The symbol with an exclamation mark, used together with text in orange, is used to indicate important learning points. This will be material crystallised from the main body of text and repeated because of its significance.

The book symbol inscribed *Intego Case Study,* together with text in light blue, is used whenever work is done on the case study which is used throughout the book. This is used to separate the details of the case study from the body of text.

The symbol of a laptop computer and compact disks used in conjunction with text in light grey is used whenever reference is made to one of our websites. The websites are sources of articles, templates, procedures and training modules that can be accessed to simplify the task of programme management.

Closing comments

Our aim was to capture and share practical lessons learnt through many years of experience on programmes. We sincerely hope that you find this book of interest and that it contributes to the successful completion of projects, programmes and portfolios in your company.

We are available to assist you with your projects and programmes through our company, Owner Team Consultation (Pty) Ltd.

Freek van Heerden
Jurie Steyn
Davida van der Walt

June 2015

This page intentionally left blank

PART 1
Setting the Scene

Chapter 1: Putting Programmes in Context

Chapter 2: The Owner Organisation

Chapter 3: A Programme Management Model

This page intentionally left blank

Chapter 1:
Putting Programmes in Context

"Project management is like juggling three balls – time, cost and quality. Programme management is like a troupe of circus performers standing in a circle, each juggling-three balls and swapping balls from time to time." — Geoff Reiss

Introduction

There are many excellent books on project management, project portfolio management and programme management. Many project managers started their careers using the concepts and methodologies taught to them by Harold Kerzner, based on his book: *Project management – a systems approach to planning scheduling and controlling,* now in its 11th edition (Kerzner, 2013).

There are also a number of different bodies of knowledge for managing projects, programmes and project portfolios, standards for project management and individualised project management qualifications. The most well-known of these are shown in Figure 1.1, together with their specific regions of application. Countries without their own guidelines, standards and certifications typically use a combination of these and other best practices.

In North America, the Project Management Institute (PMI) prevails. The PMI *Project Management Body of Knowledge (PMBOK),* fifth edition, is a collection of processes and knowledge areas accepted as best practice for the project management profession. As an internationally recognised standard it provides the fundamentals of project management, irrespective of the type of project, be it construction, software, engineering or automotive (PMI, 2013a).

PMI have separate best practice guides for programme management (PMI, 2013b) and project portfolio management (PMI, 2013c).

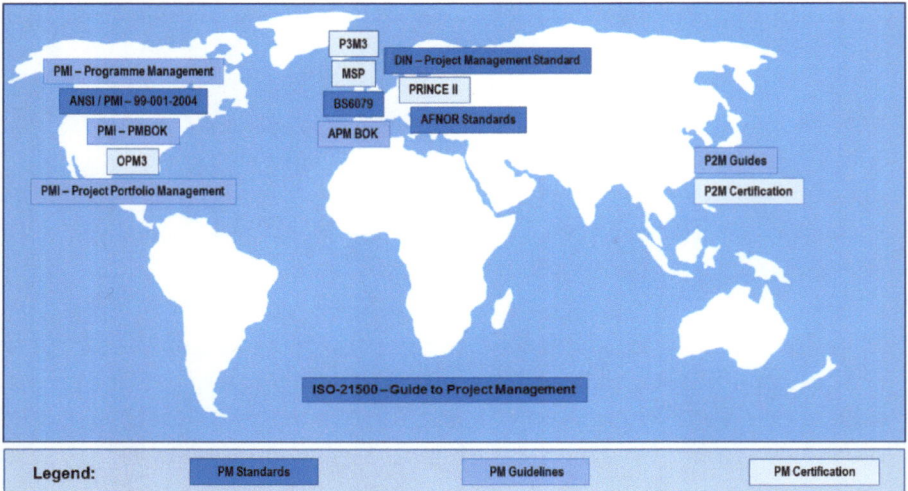

Figure 1.1: Overview of project management standards and best practice guidelines

A best-practice framework for delivering complex programmes in accordance with long term strategies is *Managing Successful Programmes (MSP)* (OGC, 2011) of the UK Office of Government Commerce (OGC). MSP® was first released in 1999 in recognition of the need for greater links between an organisation's longer-term strategy, objectives and goals and the projects being undertaken by that organisation. The emphasis is on stakeholder engagement and benefits realisation management. MSP® is independent from, but supplementary to the OGCs PRINCE2™ (an acronym for PRojects IN Controlled Environments, version 2) methodology for managing projects. In fact, Murray (2010) describes how MSP® and PRINCE2™ can be used together on programmes with several sub-projects. A useful portfolio, programme and project management maturity model was added in 2003. Both MSP® and the maturity model have been significantly revised during the past few years by expanding on the original concepts and introducing new tools and techniques.

The Association for Project Management (APM) is a professional body, registered in England for the improvement of project outcomes. The 6th edition of their project management guidelines, the *APM Body of Knowledge*, was published in 2012 and is a collaborative work based on the experience of practitioners, influencers and academics in all disciplines and sectors (APM, 2012). It provides comprehensive coverage of the disciplines of project, programme and portfolio management. These guidelines are primarily used in the UK.

The International Organization for Standardization published an international standard, *ISO 2150:2012 – Guidance on project management,* in October 2012, building on the existing bodies of knowledge of project management. The intent was to establish an overarching global body of knowledge/guideline for project management. It is to be a common platform or a reference framework for the project management community, facilitate knowledge transfer and the harmonisation of principles, vocabulary and processes in existing and future standards. This standard places projects in the context of programmes and project portfolios. It does not provide detailed guidance on the management of programmes and project portfolios, although these may be included in future revisions of, or additions to, the standard.

Learning from best practice is essential to mitigate against the risks of technology, integration, capital cost and schedule associated with projects and programmes. In this regard there are two professional organisations that can be of benefit to any project-driven organisation, namely the Construction Industry Institute (CII) and Independent Project Analysis, Incorporated (IPA). CII is a consortium of business owners, designers, engineers, construction companies, suppliers and universities formed to improve the capital project delivery process from pre-feasibility through completion and commissioning. IPA is a research organisation that specifically focuses on the drivers of capital project success. IPA has a global database of over 14,000 projects and works regularly with over 100 major companies.

However, the focus in this book is on what you, as a member of an owner team, need to do and look out for to ensure that your programme runs smoothly and meets the associated strategic objectives and benefits for the organisation. The intent is not to provide the reader with a comprehensive guide on project management, but rather a consolidation of lessons learnt and best practices we have gained by being part of programme management teams.

Chapter 1 focuses on facilitating a better understanding of what a programme is and how it differs from projects and project portfolios. Once the concept of a programme is clear, we move on to aspects of project management, programme management and portfolio management. In chapter 2 we define what owner organisations and owner teams are, in order for readers to understand the intent of this book.

Putting portfolios, programmes and projects in context

Introduction

This section on project portfolios, programmes and projects is based on an Insight Article on our website (Steyn, 2015). It is repeated here, with minor improvements, because it is essential to have a clear understanding of the terminology and interrelationship of these three matters.

See our website www.ownerteamconsult.com for our free monthly Insight Articles. Insight Articles cover a range of topics regarding project, programme and business management

A pyramid hierarchy

Distinguishing between a portfolio, a programme and a project presents a dilemma because the terms are often confused with one another. Many authors still describe a programme as a portfolio of projects. This certainly does not help to clear up the confusion. However, it is important to know the difference between project portfolios, programmes and projects because each has a special role to play. They need to be managed differently if the organisation's strategy is to be successfully transformed into reality.

A good way to avoid confusion is to think in terms of a pyramid hierarchy. Before focusing on portfolios, programmes and projects it is necessary to start with the strategy of an enterprise. The enterprise strategy presents a vision of where the organisation wants to be in future. Therefore the enterprise strategy is placed at the top of the pyramid. In organisations comprising multiple business units, the enterprise strategy has to be interpreted and incorporated in the business unit strategies. The top two layers of the pyramid thus purely focus on the strategic management component of the organisation. The pyramid structure is shown in Figure 1.2.

The business unit strategies will give rise to strategic management plans, which in turn will generate a portfolio of programmes and/or projects to be implemented per business unit, prioritised according to business objectives and needs and within the resource constraints. As you go up the pyramid from the bottom (project to programme to portfolio) then the budget, life expectancy, complexity and interdependencies all become greater. Programmes can also reach across several or all the business units in the enterprise.

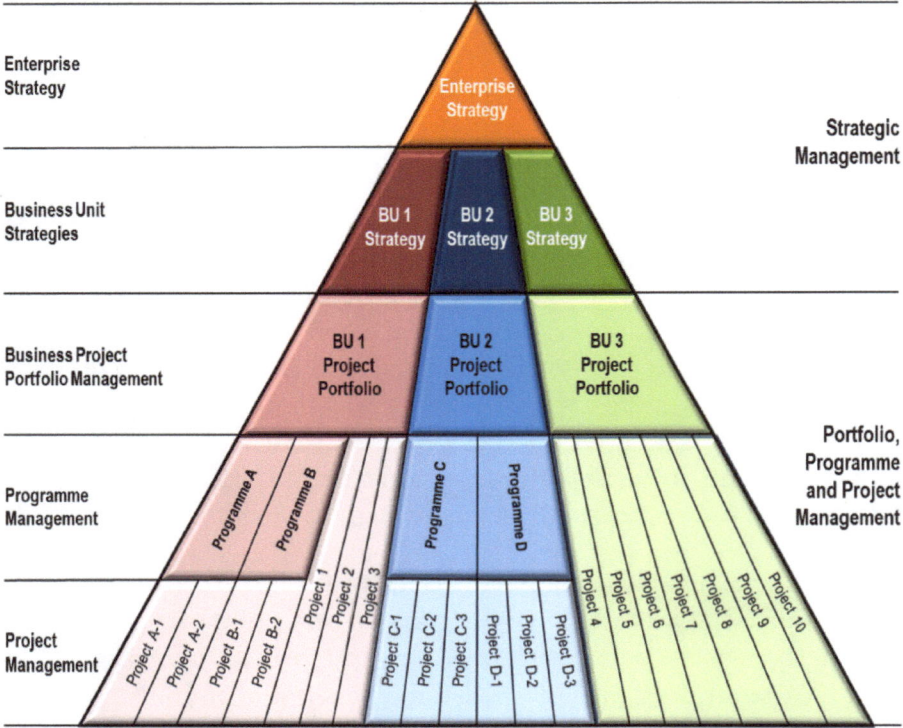

Figure 1.2: Putting portfolios, programmes and projects in context

We now need to consider the definitions of a project, a programme and a project portfolio to highlight the differences. Definitions from the US Project Management Institute (PMI) and the UK Office of Government Commerce (OGC) will be used for this discussion:

At the lowest level of the pyramid there are projects. These projects can form part of a programme, or not. A project is:

- a temporary endeavour undertaken to create a unique product, service, or result (PMI, 2013a), or alternatively;

- a unique set of co-ordinated activities, with definite starting and finishing points, undertaken by an individual or team to meet specific objectives within defined time, cost and performance parameters as specified in the business case (OGC, 2010).

Some projects form part of programmes, whilst others do not. This is reflected in the second level from the bottom of the pyramid. A programme is:

- a group of related projects managed in a coordinated way to obtain benefits and control not available from managing them individually (PMI, 2013b), or alternatively;

- a collection of projects that are linked together by a business need and clearly defined benefits. (OGC, 2010).

Programmes thus consist of multiple projects. During the lifespan of a programme, projects can be added to the programme or removed as the overall strategy becomes clearer. Overall expenditure on programmes is typically significantly greater than that for projects. These efforts will also consume significant amounts of funding which can translate into hard choices about whether to continue or discontinue programmes or certain aspects of them. An example of a programme is an IT programme, designed to improve customer service, which contains an internet upgrade project, invoicing and customer complaints projects. A programme is therefore likely to involve a number of different departments or functions within the organisation and can span several years.

Programmes and individual projects (that do not form part of programmes) roll up into portfolios as shown in the third level from the bottom of the pyramid. The correct term to use is 'project portfolio', so as not to confuse it with financial investment portfolios. A project portfolio is:

- a collection of projects and/or programmes and other work that is grouped together to facilitate the effective management of that work to meet strategic business objectives (PMI, 2013c), or alternatively;

- a collection of programmes or projects that define the totality of the organisation's investment in change to facilitate strategic business objectives (OGC, 2010).

The programmes and projects in an enterprise or business unit project portfolio may not be related, except to the extent that they are identified, prioritised and authorised with a view towards achieving a set of strategic business objectives. For the effective management of projects and optimal use of joint resources, several projects may be grouped into sensible categories (e.g. location, business area, types of projects) under a specific project manager. Some organisations refer to such a manager-of-projects as a portfolio manager, but this should rather be a project portfolio manager.

The definitions presented above are all succinct and to the point. Although the emphasis may be slightly different between the PMI and OGC definitions, the interpretation thereof is very similar. Other definitions can be found in the literature, but the ones presented here have wide support in the project environment.

Enterprise-wide initiatives

The way in which Figure 1.2 is presented, reflects an enterprise comprising three business units, each with its own strategic business unit objectives and project portfolios. In such a case, how will corporate or enterprise-wide initiatives or programmes be handled and should there be an enterprise project portfolio? The answer is yes, but it need not be a permanent organisation structure. Programme and project specialists can be seconded from other business units for the duration of the enterprise-wide programmes.

We will consider three enterprise-wide initiatives or programmes. Each of the three enterprise programmes consists of a number of projects as detailed in Figure 1.3.

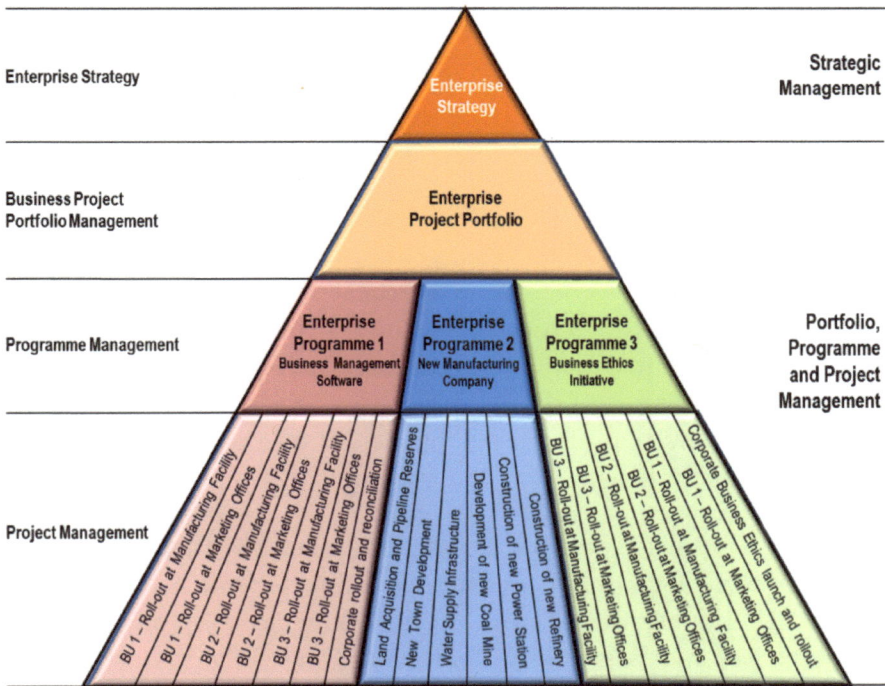

Figure 1.3: Enterprise project portfolios and programmes

Whereas enterprise programmes 1 and 3 impact on all the existing business units and structures in the organisation, programme 2 does not. The focus of programme 2 is to establish a new business unit with a totally new operating facility and product marketing department in an undeveloped area. Enterprise programmes thus do not only impact existing business units, but may actually establish new business units for organisation growth.

Three examples of enterprise-wide initiatives or programmes are given in Figure 1.3, namely the roll-out of new business management software to the entire organisation; the establishment of a new manufacturing company, and; the roll-out of a business ethics initiative for the organisation. An enterprise project portfolio function is also shown for prioritising and managing enterprise-wide programmes.

Project management

Projects and key attributes

A project can be anything, from the construction of a new building, the implementation of a more powerful computer software system, a major staff reorganisation, a merger, acquisition or disposal within a business, the roll-out of a new product or service to the construction of a major new manufacturing facility. It is therefore difficult to provide one simple definition of a project.

We've already discussed the Project Management Institute and the UK Office of Government Commerce definitions of a project. The Oxford Dictionary definition, describes a project as "an enterprise carefully planned to achieve a particular aim". The Association for Project Management (APM, 2006) adds the idea that a project has a limited lifespan by defining a project as "a unique transient endeavour undertaken to achieve a desired outcome."

Using these and other definitions, it is possible to identify the key characteristics of a project:

- A project is finite - activities have a defined start and finish;
- Every project is unique and has its own challenges;

- Project deliverables are designed to meet specific business objectives;

- Coordinated activities are undertaken to create the deliverables, and;

- The project requires specific resources.

Defining project management

The concept of a project should now be well entrenched, namely unique and transient endeavours to achieve a desired outcome for the organisation. The next step is to consider what is meant by project management.

We define project management as the process by which projects are defined, planned, monitored, controlled and delivered. It involves the application of knowledge, skills, techniques and resources to execute projects effectively and efficiently. Project management is a strategic competency for organisations, enabling them to tie project results to business objectives.

Project management is the process by which projects are defined, planned, monitored, controlled and delivered. It involves the application of knowledge, skills, techniques and resources to execute projects effectively and efficiently.

Project size and complexity

The total monetary value of a project is obviously important and it can be used to categorise projects. The project descriptors and typical monetary value range (2015 dollars) for each is shown in

Table 1.1. Here we identify seven categories of projects, from enhancement projects, through small, medium, large, very large, megaprojects and gigaprojects. Very large projects can also be referred to as super projects. The term gigaproject has only been used for the past few years, but the use thereof is expected to grow in future as more projects enter the range of greater than $10 billion.

Table 1.1: Categorising projects according to monetary value

Descriptor	Monetary Value (USD)
Enhancement projects	< $250 000
Small projects	$250 000 - $1 million
Medium projects	$1 million - $10 million
Large projects	$10 million - $100 million
Very large projects (Super projects)	$100 million - $1 billion
Megaprojects	$1 billion - $10 billion
Gigaprojects	$10 billion - $100 billion

The category of project cannot only be based on project size based on the monetary value thereof. Most project managers define a project's category based on some or all of the following parameters:

• Total financial resources required;

• Size and composition of the project team;

• Number and size of deliverables to be produced;

• Complexity of deliverables to be produced, and;

• Timeframes involved in delivery.

As the project size increases, the complexity of the project will increase as well, as illustrated in Figure 1.4. However, in almost all cases the category of project will be defined by its budgeted expenditure.

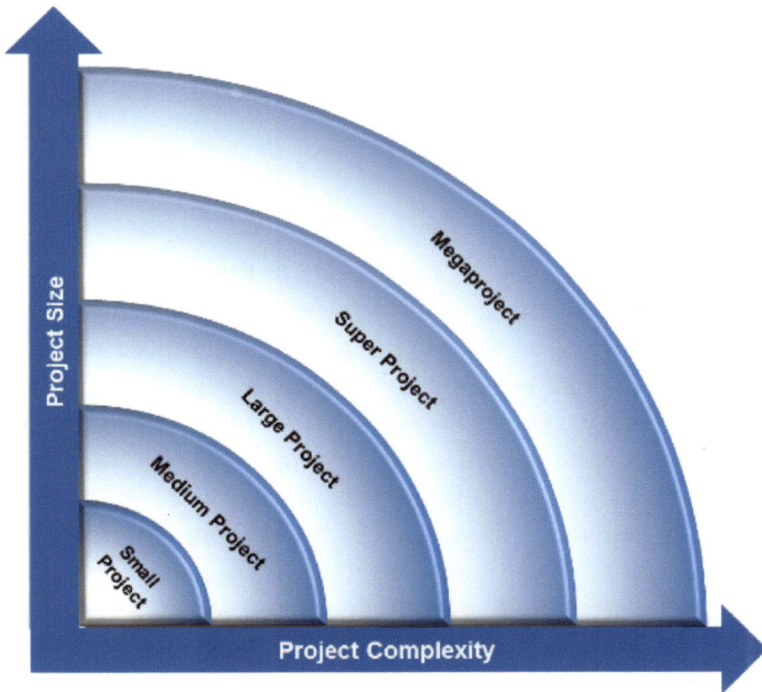

Project Size

Megaproject

Super Project

Large Project

Medium Project

Small Project

Project Complexity

Figure 1.4: Project size versus complexity

Buttrick (2010) maintains that some projects are simply too large to manage as a single entity and recommends that a programme approach is followed. His concept is that the organisation structure for programmes is one where a programme manager is supported by multiple project managers, each with his own project team. In terms of this statement, it is typically the large projects, super projects, megaprojects and gigaprojects that would be handled using a programme management approach.

It is typically the large projects, super projects, megaprojects and gigaprojects that would be handled using a programme management approach.

Programme management

Definition

The concept of programmes has emerged over the past fifteen years. Many large- and megaprojects today are managed as programmes, rather than as projects. Managing people and organisational issues is at the centre of programme management. The ultimate measure of success is whether the business achieves its objective, whereas each project within the programme has a very specific deliverable against a timeline and cost. Programme management is thus more concerned with managing the bigger picture. This picture is often unclear and develops over time in contrast to project management that has greater definition and is implemented over a shorter time span.

Programme management, in addition, involves setting and reviewing objectives in a coordinated and integrated fashion, coordinating activities across projects, and overseeing the integration and reuse of interim work products and results. A programme requires a programme director whose major responsibility is to ensure that the work effort achieves the outcome specified in the business strategy through proper integration of the individual projects. This person spends more time and effort on integration activities, negotiating changes in plans, and communicating than on the other project management activities we described (Hanford, 2004).

We define programme management as the co-ordinated organisation, direction and implementation of a group of related projects and activities that together achieve a specific outcome and realise benefits that are of strategic importance.

Programme management is the co-ordinated organisation, direction and implementation of a group of related projects and activities that together achieve a specific outcome and realise benefits that are of strategic importance.

Critical success factors for programme management

Critical success factors are typically identified to enable tracking the value of programme outputs. Ideally, success factors should enable the accurate monitoring of the few fundamental aspects of a programme that must be done well to achieve the strategic objectives of the programme. Success factors are those inputs that lead either directly or indirectly to the success of a project or programme, whereas success criteria are those measures against which the success or failure of a project/programme will be judged. Identifying and communicating the critical success factors ensures that everyone in the programme team is focused.

Shehu and Akintoye (2009) identified five principal factors for programme success under which they grouped the critical success factors as obtained from their study on the UK construction industry. The principal factors and the critical success factors under each of these are listed in order of significance in Table 1.2.

This is an extensive list of critical success factors for programmes in the construction industry. However, for programmes implemented in an owner organisation, the accessibility to management is

17

considered to have too low a priority. Top management support for a programme, in the person of an effective programme sponsor, is essential. Effective executive sponsorship is considered one of the most important ingredients required for successful programme and project execution in owner organisations.

Table 1.2: Principal and critical success factors for a programme

Principal Success Factors	Critical Success Factors
Programme Coordination	• Cross-discipline coordination • Cross-discipline problem solving • Cross-project coordination • Effective integrated risk management • Effective change management • Effective management of transition • Effective communication • Proper coordination of projects
Programme Priority Focus	• Effective planning • Establishing programme priorities • Effective performance management
Programme Vision	• Clarity and consistency of vision • Understanding the stakeholders' attitude • Clear benefits target • Strategic focus on programme
Programme Strategy	• Simplicity of programmes • Ease of use of tools and techniques • Effective quality management • Accessibility to management infrastructure
Programme Planning	• Effective time management • Effective budgeting • Smooth handover to business operations on completion

Why are programmes important?

Programmes are similar to projects in the context of helping organisations grow, maintaining competitive advantage and/or helping ensure compliance with legal, environmental and social requirements. The focus with a programme is more on the strategic business objectives and not only on cost, schedule and quality as with a project. The impact of programmes not meeting an organisation's strategic targets could easily be a factor of 10, or more, higher than that of a project.

Seeing that a programme is a "group of related projects managed in a coordinated way to obtain benefits and control not available from managing them individually (PMI, 2013b)", the risk is there that the failure of only one of the related projects could undermine a programme to such an extent that the strategic programme objectives are not achieved. Therefore, for a programme to be successful, all the related projects in the programme have to be successful.

For a programme to be successful all the related projects in the programme have to be successful and the strategic business objectives of the programme have to be met.

Typical Programme Management Challenges

Due to the complexities outlined above, some of the frequent programme management challenges are:

- Conflict in resource allocation. Different, but related projects share the same resources. This challenge can be overcome by an experienced programme director, provided the required funding is available;

- Different projects have different and changing business priorities and therefore different levels of commitment. This is realised in the resource allocation: Over-allocation in the committed project, under-allocation in the lower priority projects;

- Scope changes of the individual projects because of internal organisational issues or external market driven issues, in order to ensure that the overall business objectives are met;

- Conflicting demands from the different project sponsors of the various sub-projects in the programme should not apply if a single, strong, executive sponsor for the programme is appointed. It helps to use the affected business owners as sponsors for the sub-projects in their areas, and;

- The various sub-project sponsors and/or business owners have their own view on requirements and goals of the overall programme (and strategy). Sub-project business owners tend to overemphasize the importance of their business unit's objectives. Again, this should not apply if a single, strong, executive sponsor for the programme is appointed.

When is a programme approach applicable?

Programmes should be seen as much more than just the technical management of projects. Programmes function on a much larger scale than most projects. The outcome of a programme can have a significant impact upon financial results, product viability and business sustainability. A programme can be the result of a business objective such as to improve efficiency of operating facilities by 20% over the next 15 years. Due to the nature of such a programme, the scope tends to be less clear than a normal project. Definition of the programme scope develops over time and this, in turn, leads to uncertainty in cost and schedule.

Programmes are particularly suited to enable the achievement of a specific business objective, for example whether it is to:

- Grow the business through the introduction of a new product into the market;

- Introduce new integrated business management software throughout the organisation;

- Achieve legislative and social targets such as the reduction of the environmental footprint of the organisation;

- Entrench new business governance procedures and an ethical management policy in the organisation;

- Improve the overall business competitiveness by cost reduction, or;

- Establish a new facility in areas without appropriate infrastructure.

The trend at the moment is that projects are generally much larger, more complex and over much longer duration than the traditional three to five year lifespan. Utilising programme management concepts offer the benefit of total oversight, accountability and continuity that is so desperately required to ensure successful completion.

A trap that must be prevented is to try and expand the remit of a programme to such an extent that it becomes all inclusive of the total business strategy. This implies managing programmes as project portfolios. Programmes, like projects, still need to be well defined with clear boundaries and a single business objective.

According to Haughey (2001), a group of related projects not managed as a programme is likely to run off course and fail to achieve the desired outcome. It is therefore important that programmes are run within a framework that ensures there is a focus on the overall strategic objectives.

A group of related projects not managed as a programme is likely to run off course and fail to achieve the desired outcome (Haughey, 2001).

Project portfolio management

For the purpose of this book, we will use the definition of project portfolio management as given in the 6th edition of the APM guidelines (APM, 2012). Thus, project portfolio management is the selection, prioritisation and control of an organisation's projects and programmes in line with its strategic objectives and capacity to deliver. The goal is to balance change initiatives and business-as-usual while optimising return on investment.

Project portfolio management is the selection, prioritisation and control of an organisation's projects and programmes in line with its strategic objectives and capacity to deliver.

Concluding remarks

In this chapter, the concept of projects, programmes and project portfolios was considered to highlight the unique aspects of each. Programme management was defined and discussed and the concept of the owner organisation was mentioned.

In the following two chapters of Part 1, the owner organisation and its role are discussed in detail and a programme management model is introduced. We also introduce a case study that is used as a worked example throughout the book.

In Parts 2 and 3, the various elements of the programme management model are discussed in detail. Practical examples are given at each step, using the case study as the vehicle therefore.

Chapter 1

This page intentionally left blank

Chapter 2:
The Owner Organisation

"Whenever we seek to avoid the responsibility for our own behaviour, we do so by attempting to give that responsibility to some other individual or organisation or entity. But this means we then give away our power to that entity." - M. Scott Peck

Introduction

Owner organisations play a very important role in any project or programme. They will own and use the results of the projects and programmes they initiate. Successful projects and programmes will benefit them and their stakeholders directly. On the other hand, unsuccessful projects and programmes may lead to their demise.

The owner organisation should therefore maintain the power to control the outcome of their projects and programmes, and not readily give that power away to their engineering and managing contractors. The guidelines presented in this chapter, and remainder of this book, should assist owner organisations to retain power and positively impact the probability of success of their projects and programmes.

In this chapter we firstly consider the key players in a project and programme environment, and discuss the profile of a typical programme manager. Secondly, the role of the owner organisation, and how it differs from that of engineering contractors, is described. Typical structures for owner project management teams and owner programme management teams are presented. Lastly, we briefly discuss the role and importance of an executive programme sponsor.

Key players within programmes and projects

Roles in the project/programme management process

A successful project requires a wide range of stakeholders to cooperate and work together. The nature of their roles will depend upon the scale of the organisation, the type of projects, and the size of the project portfolio or programme.

The following are the potential roles which may be part of the project/programme management process. There are likely to be variations between organisations and it is useful to recognise that these are roles and may not necessarily be individual posts. In some cases, a role may be combined into part of a person's larger job; in smaller organisations, one person may have a number of roles.

- **Approval body:** The approval body is the group for whom the project is being undertaken. They appoint the project/programme sponsor and provide financing;

- **Project sponsor:** This is the individual who is accountable to the approval body. The project sponsor is usually an executive manager responsible for the business charter, holding the project budget and delivering the business benefits;

- **Steering Committee:** The steering committee is a body of experts appointed by the sponsor to advise and guide the sponsor in decision making;

- **Contractors:** Contractors are the groups who provide the expertise to do the actual work on the project (i.e. they will be designing and building the outcome). This may be an in-house or contracted service;

- **Programme manager:** This is the person for managing a programme consisting of a number of different projects, each with its own project manager, and is accountable to the project sponsor;

- **Project manager:** This is the person responsible for managing a specific project and is accountable to the programme manager;

- **Business owner:** Business owners specify operational requirements and are also the people who will ultimately accept the end product from the project or programme. User responsibilities include identifying the project requirements, stating project constraints and testing;

- **Stakeholders:** This is a person, group or organisation that is affected by or has an interest in the activities of the project, and;

- **Project team:** The project team is responsible to the project manager for undertaking tasks and managing risks within the culture and constraints of the project.

Profile of a programme manager

According to Brown (2008), a programme manager should first and foremost be a leader. He continues that the programme manager's main leadership duty is to turn chaos into clarity for the team. People need clear direction and circumstances that allow them to be successful. The programme manager must establish such direction both within and outside the organisation through a variety of means. Additionally, the programme manager may have to accept calculated risk when he or she is unable to obtain clarity from top management of the organisation and then define clarity in his or her own terms. Accepting chaos, allowing chaos to exist, or passing down chaos all signal a lack of integrity and this does not create a culture conducive to successful programmes.

The programme manager is responsible for creating the environment and culture within which the project manager executes a project. The degree of the programme manager's direct control of that environment and culture can vary, but through direct authority or organisational influence he or she is responsible for establishing the framework in which the project manager operates.

The project manager is judged on the triple constraint of schedule, cost and quality (the three cornerstones of the scope of the project). The programme manager also is judged on these three elements but at a level that is cumulative for all the projects and operations within the programme. This aggregation of responsibilities for a variety of projects and operations means the programme manager must make frequent trade-offs between business targets and project/operational performance.

Programme management decisions are both tactical and strategic in nature. The strategy aspects of these decisions must consider multidimensional impacts beyond the near-term delivery dates of the project. Conversely, the project manager is expected to deliver projects within the boundaries and framework established by the programme manager. Typically, the project manager should be more delivery and execution focused whereas the programme manager has to also be concerned with the overall health, effectiveness and strategic focus of the programme over the long term.

The project manager should be delivery and execution focused, whereas the programme manager also has to be concerned with the overall health, effectiveness and strategic focus of the programme over the long term.

The Owner Organisation

Introduction

This section on the owner organisation has been published before as an Insight Article on our website (Steyn & Lourens, 2014). It is repeated here, with minor improvements, because it is imperative that the reader has a clear understanding regarding the owner organisation and the owner project management team.

See our website www.ownerteamconsult.com for our free monthly Insight Articles. You can subscribe on the website to have these articles e-mailed directly to your mailbox. Insight Articles cover a range of topics regarding project, programme and business management

Definition

The owner organisation, as the name implies, is simply the organisation or entity that will own, operate and benefit from the result of successful project or programme execution. The owner organisation executes projects and programmes to meet specific business objectives.

In the case of an unsuccessful project or programme, it is the organisation that will own the (unproductive) assets and carry the consequences in terms of opportunities, income and reputation lost.

Owner Project Management Team

The owner organisation wishing to implement a project can either do it in-house or engage one or more contractors do it on its behalf. For large projects (projects of between $10 million and $100 million) or greater, the ideal approach is to establish an owner project management team, supported by a managing contractor. The managing contractor typically supplements the owner organisations skill set in managing the overall programme.

Research on large project outcomes is very clear on the crucial role played by strong, fully staffed owner project management teams (Merrow, 2011). The role of the owner team is to bring all facets of the owner organisation's requirements to the table, especially during the scope development phase. Merrow (2011) maintains that projects fail when they experience significant changes after the completion of front-end engineering design. If it is not entirely clear what the owner organisation requires, major changes are inevitable and with the changes will come inevitable failure.

The owner organisation and owner project management team is structured as shown in Figure 2.1. The project is executed on behalf of the owner as represented by the approval body (normally a board of directors) and the project sponsor, with the assistance of a project steering committee. The owner project team executes the project on behalf of the owner and is represented by the triangle that comprises the business manager, the engineering manager (and team), operations manager (and team) and the project manager (and team). The project manager is the overall leader and the accountable person, as depicted by his central position.

The entities indicated in blue in Figure 2.1 are all owner organisation personnel and need to take a long-term perspective on the project to ensure that a sustainable business is established. This is in contrast to the engineering and other contractors who will be managed by the owner team (as shown in the brown block) and who will have a more short-term orientation, focused on delivering the product as specified by the owner.

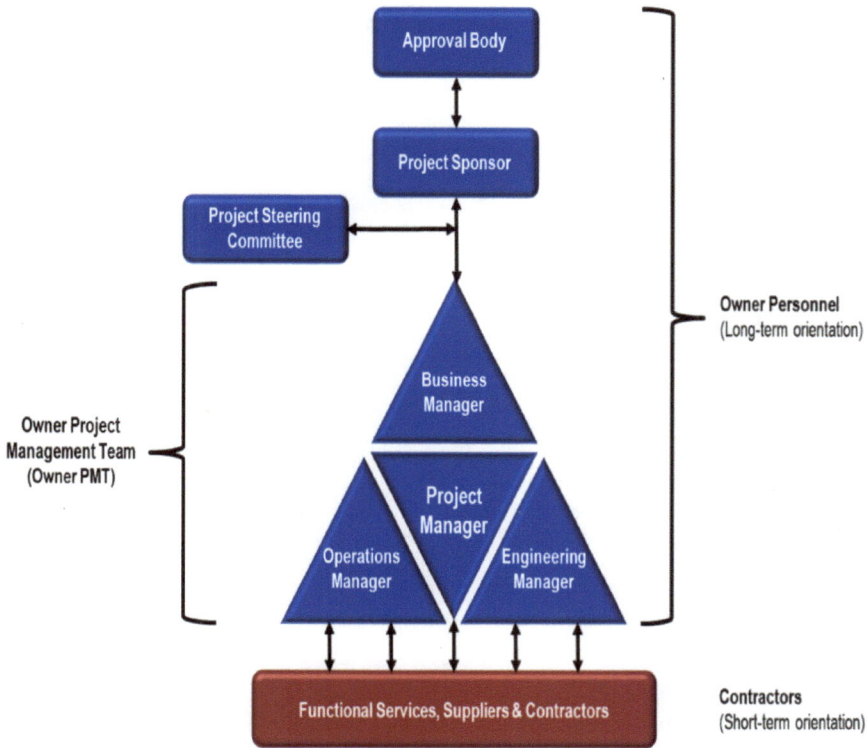

Figure 2.1: **The owner organisation and owner project management team**

Owner Programme Management Teams

The organisation structure becomes much more complex for large, super-, mega- and gigaprojects which are handled as programmes. Since programme management is the coordinated management of related projects, the recommended structure for programmes involves the structure shown in Figure 2.1 for all the related projects in the programme under an owner programme management team, led by a programme director (see Figure 2.2).

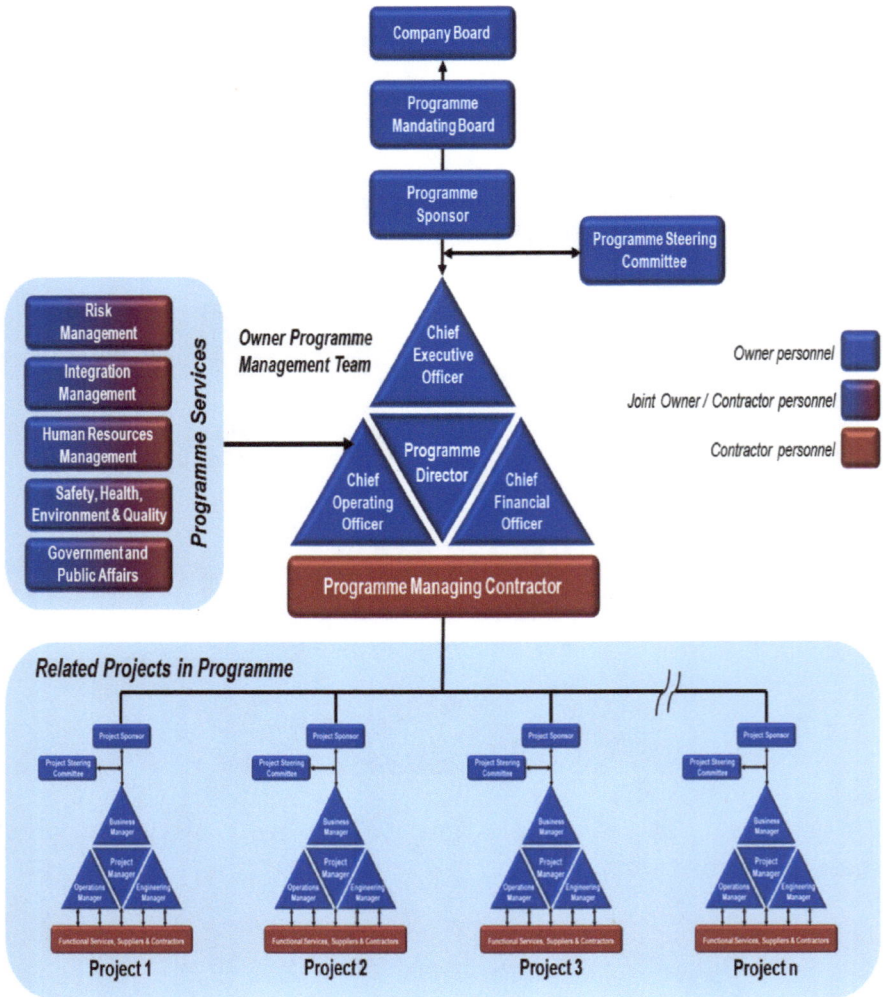

Figure 2.2: The owner programme management structure with related project management teams

The owner programme management team is structured as shown in Figure 2.2. The programme is executed on behalf of the owner as represented by the company board, the programme mandating board and the programme sponsor, assisted by a programme steering committee.

Figure 2.2 also shows the joint programme services such as integration management, human resources management, safety, health, environmental and quality management and programme specific government and public affairs. These programme services should preferentially be staffed with owner personnel, but is typically a combination of owner and contractor personnel. Refer to Chapter 6 on shaping the programme organisation for more detail.

Owner organisations and contractors have different roles

As alluded to in the previous section, the owner organisation and the contractors have different roles and perspectives. The owner organisation will own and operate the new facilities and hardware for business gain, whereas the contractor owns systems and tools and sells man-hours for business gain. Contractors are indispensable for the execution of large, super-, mega- and gigaprojects, but can never substitute for owner project teams.

Owner organisations have to:

• Have a long-term, strategic view;

• Distinguish what it is about their particular company that generates unique competitive advantage;

• Understand the total organisational supply chain from raw material supply to the actual production operations as well as final product distribution, and;

• Translate their business understanding into a productive asset, not just a 'good project'.

Managing contractors cannot perform this work on behalf of the owner organisation because they:

- Have a short-term view relative to the owner's perspective, typically only up to the end of the programme;

- Do not understand the business of the organisations they serve;

- Do not understand the owner organisations' competitive advantage (or disadvantage);

- Are inexperienced in production and operations, and;

- Work for their own shareholders, not the shareholders of the owner organisations.

The owner is the only party that can develop the specifications of a new product/facility to meet an identified business need. Secondly, the owner needs to ensure that the product/facility is developed according to the specifications and then used effectively for business gain.

According to Read (2004), both the project managers of the owner organisation and the managing contractor are essential elements in the successful execution of a project. He maintains that their roles are not only different, but also complementary.

Project and Programme Sponsors

In order to achieve success, a programme needs a strong, full-time executive sponsor that is accountable for the business outcome. What should the perfect project and programme sponsor look like? According to Kloppenborg et al (2006), project sponsor characteristics, as cited most frequently in the literature, include:

- Appropriate seniority and power in the organisation;

- Political knowledge and savvy;

- Ability and willingness to make connections between the project/programme and the organisation;

- Courage and willingness to stand up and defend the project or programme;

- Ability to motivate the project team and provide ad hoc support as required;

- Willingness to partner with the project team and project manager or programme director;

- Excellent communication skills;

- Personally compatible with other key members of the owner project team, and;

- Ability and willingness to challenge the project and provide objectivity.

Concluding Remarks

We now know what an owner organisation is and why it is important to retain the power to steer a project or programme to the desired outcome. For an owner organisation this means a significant investment in resources to properly set up a programme management office and programme management teams. The importance of a suitably qualified executive sponsor for every programme cannot be overemphasised.

We introduce a programme management model for owner teams in the next chapter that forms the basis for the remainder of this book.

This page intentionally left blank

Chapter 3:
A Programme Management Model

"We don't care about what you did yesterday—we care about what you're going to do tomorrow." — *Cory Doctorow*

Introduction

The concepts of programme management, the owner organisation and the owner programme management team were discussed at length in Chapters 1 and 2.

As part of setting the scene, in this third chapter, a programme management model for owner teams is introduced, which will be expanded on in the following chapters. The case study that will be used throughout the book to demonstrate how to put theory into practice is also introduced.

A programme management model

Business versus technology strategy

In any business where technology is essential to the business model, the business strategy is normally underpinned by a technology strategy. The business strategy provides the business pressure to produce and distribute what the market demands. The technology strategy, on the other hand, dictates what is possible and determines what can be produced by the business in light of technology trends and advancements. This constructive tension helps the business to achieve best results through consideration of what the market wants, but also influences what they want through what technology can provide.

From this highly interactive process between business strategy and technology strategy, programmes are born. A programme therefore comprises a number of projects required to achieve the specific business strategic objective. These projects may (and is very likely to) be executed according to different timelines in terms of their individual development and roll-out. A programme manager will be required to develop and manage the interrelated projects in order to achieve the overall objective. Figure 3.1 shows how programme management helps realise the business strategy.

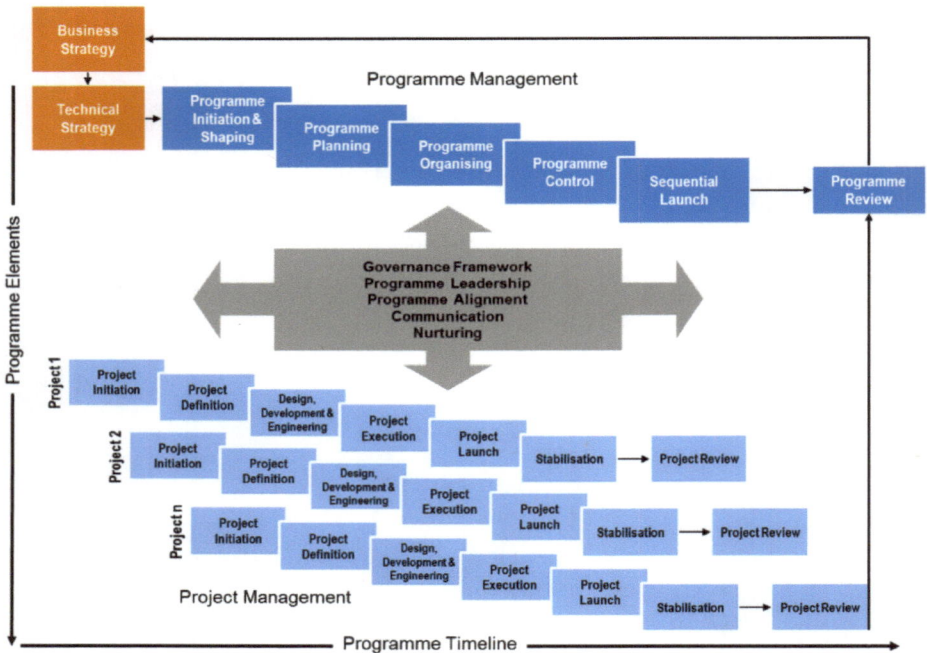

Figure 3.1: Programme management helps implement business strategy

Traditionally during the project initiation phase, a business opportunity supporting a specific strategic objective is identified and defined as a project or sub-project up to a point where the project manager can assume control. This typically occurs after completion of the business idea generation and at least the project definition (prefeasibility) stage. The project manager then manages the technical execution of the project up to ready for market (ready for launch) when the marketing or operations team takes ownership to establish the running concern as an ongoing business venture.

The concern with this approach is that it is, at best, 'fragmented' in terms of whom is responsible for the different phases of the project or programme, as illustrated in Figure 3.2. Responsibility is either shared or passes from one department to the other along the different project stages. Shared responsibilities or a frequent change in the responsible party is a sure recipe for inefficiency and potential project schedule slippage.

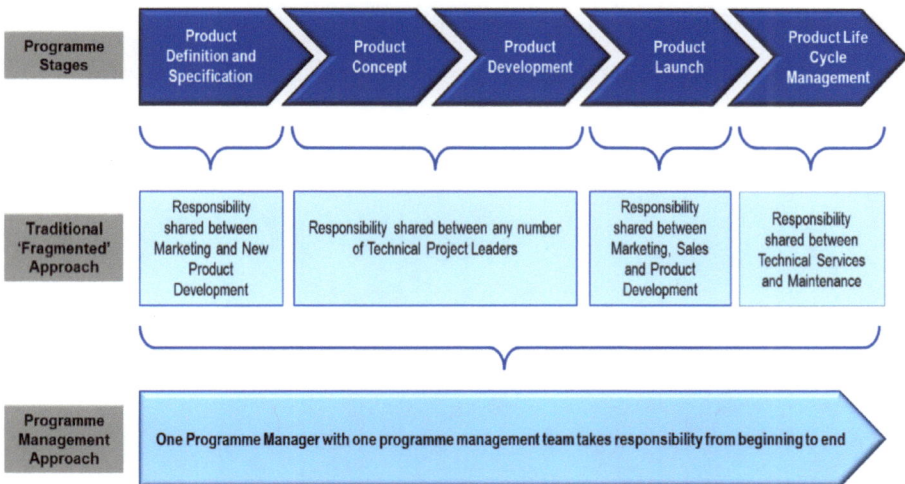

Programme Stages	Product Definition and Specification	Product Concept	Product Development	Product Launch	Product Life Cycle Management
Traditional 'Fragmented' Approach	Responsibility shared between Marketing and New Product Development	Responsibility shared between any number of Technical Project Leaders		Responsibility shared between Marketing, Sales and Product Development	Responsibility shared between Technical Services and Maintenance
Programme Management Approach	One Programme Manager with one programme management team takes responsibility from beginning to end				

Figure 3.2: Traditional vs programme management approach

It has been shown that inclusion of all the activities from idea stage to the final running business under a programme manager, greatly improves the overall chance for success of the initiative (Deschamps & Nayak, 1995). This implies that a single programme manager with one programme management team remains responsible throughout. This concept is included in Figure 3.1 and also illustrated in Figure 3.2.

The approach of utilising a programme manager across the total development cycle as well as managing the interdependent projects towards an overall business objective has various benefits, namely:

- An improved concept development and a mechanism to maintain concept integrity;

- Significantly improved communications;

- A clearer structure and leadership to ensure the programme development proceeds through the initial uncertain and often recycling development phase;

- A strong degree of accountability;

- An opportunity to get final feedback from end users, and;

- An opportunity to enrich the skills of all members of the team.

Including all the activities of a programme from idea stage to the final running business under a programme manager, greatly improves the overall chance for success of the initiative.

Managing related projects as a programme also has the benefit of being able to understand the overall schedule and cost as progress is being made. In addition, it offers the opportunity to understand, explore and manage the risks and opportunities between the different projects within the programme. What makes a programme different is that it lends itself to the exploitation of opportunities for synergism between the projects.

The programme management model

There are several programme management models available in the literature and they all have merit. By presenting a new programme management model, we run the risk of proposing simply another model with limited interest and application. However, the majority of the model we propose has been evaluated and utilised with success on a number of programmes and megaprojects for an owner organisation. Further additions to the model are firmly based on accepted project management theory.

Many readers will be familiar with the classic management model of Louis Allen that categorises management activities in four categories (LAI, 2009), namely:

- **Leading:** Leadership refers to using influence to motivate team members to achieve a common vision;

- **Planning:** Planning is required to select appropriate goals and the best way to achieve them i.e. to plan that the right activities are done by the right people at the right time;

- **Organising:** Organising refers to assigning responsibility for task accomplishment i.e. the right resources need to be mobilised and organised to execute the job, and;

- **Controlling:** Control mechanisms are also put in place to monitor activities and take corrective actions to ensure that the team achieves what they set out to accomplish.

We have applied the principles of the Louis Allen management model in the context of programme management, adapted and expanded it to suit the specific requirements of a complex programme with a time span of several years. The specific point of view was from the perspective of an owner organisation for which the programme was being done. Our revised programme management model is shown in Figure 3.3.

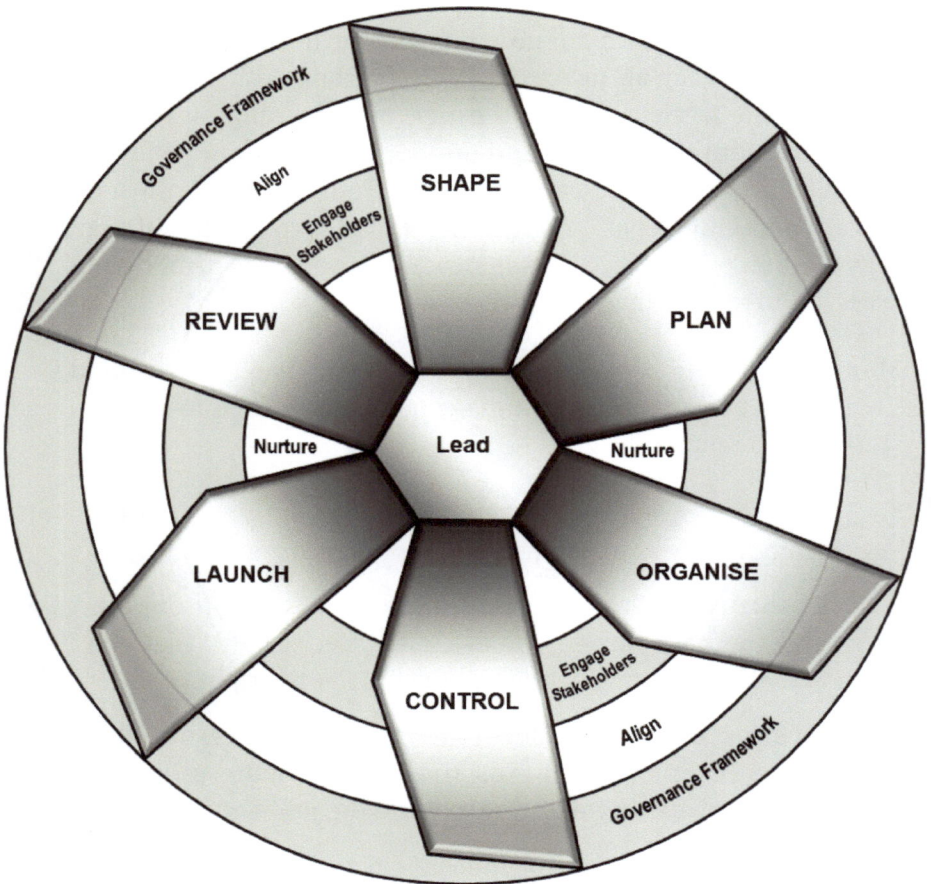

Figure 3.3: A programme management model for owner teams

As can be seen from Figure 3.3, the model essentially comprises six phases depicted as a six-bladed propeller, superimposed on a series of concentric rings (a target) covering what we refer to as the programme lifecycle essentials.

Focusing firstly on the six-bladed propeller shape, this covers the programme management processes of shaping, planning, organising, controlling, launching and reviewing. Each of these has a full chapter devoted to it in the book, but is described very briefly below:

- **Shaping:** The first step in the model is shaping the programme. This implies developing a clear business intent which provides the programme of projects with a common goal, as well a definition of a scope that could achieve the business intent;

- **Planning:** Planning is done to ensure the right things are done at the right time by the right resources;

- **Organising:** Set up the programme management team and assign responsibilities to develop and execute the plan;

- **Controlling:** Control mechanisms are put in place which includes lagging and leading indicators which will highlight any deviation from plan to allow for corrective action to be taken to ensure that programme objectives are met;

- **Launching:** Sub-projects comprising the programme need to be sequentially launched in such a manner that the overall deliverables of the programme are met, and;

- **Reviewing:** Programme review is a process that is initiated when managers and other stakeholders pause to assess how a programme has performed during a given period of time. A programme review is an integral part of the programme cycle and helps ensure continual improvement in the programme management process.

In contrast to a single project, even a megaproject, where the project phases are discrete and have clearly defined deliverables and gate criteria, the phases of a programme only depict the progression from idea to a fully-fledged programme and eventual closing off. Programme phases tend to overlap, for example: programme shaping can still be attended to while the planning work is started in earnest. As soon as planning has progressed reasonably far, the organising activities can start, followed shortly thereafter by control mechanisms. Control mechanisms need to be in place because the early projects need to be launched.

The work associated with a programme is also non-linear in nature: planning can lead to updates required in the programme shaping, or as more detail is developed during the organising phase, planning is developed further that may again lead to organising updates. This interactivity between the phases makes the management of a programme more complex, requiring both good leadership and good management from the programme leadership team. However, focus still needs to be kept on each of the phases to develop the phase fully and drive the activities to completion. The shaping phase should be closed off when the programme is sufficiently developed to meet the business objectives as stated. If this is not actively attended to, the programme could develop a life of its own, with more and more business objectives and projects being added indiscriminately.

The concentric rings cover the lifecycle essentials such as leadership, nurturing, alignment and stakeholder engagement, all within a well-defined governance framework. Again, each of these has a full chapter devoted to it in the book, but is described very briefly below:

- **Leading:** Clear leadership is essential, embodied by an executive programme sponsor who has to ensure at all times that the programme is heading in the right business direction;

- **Nurturing:** Because programmes are typically executed over an extended timeline, nurturing of people working on the programme and team wellness become critical focus areas. This includes planning for rotation and ensuring that the long-term careers of team members are catered for;

- **Stakeholder engagement:** Continuous sharing of information in such a complex system is a non-negotiable to all interested and affected parties. Communication is an essential part of team interaction. A sound communication plan and reporting system should be developed;

- **Aligning:** Even though much effort goes into initial alignment of the programme team, team alignment is a process which continues throughout the lifecycle of the programme, and;

- **Governance framework:** An owner organisation would typically have a formal ethics policy and governance procedures in place. This should be extended to also cover governance of the project environment to ensure that all dealings and interactions are such that it would not negatively impact on the reputation of the organisation.

The model is deliberately shown to be a continuous process as can be seen by the cyclical nature thereof. As the programme progresses, new business requirements may affect the programme shape, re-alignment will be required, leadership needs to ensure that everyone understands the changes and rationale behind the changes, plans need to be adjusted, resourcing needs to be reviewed, and control mechanisms confirmed. Throughout this process, communication is an essential part of team interaction. Care should be taken to ensure team wellness through nurturing of the members.

In this book we use this model as the basis for our discussions. Each of these elements is reviewed and the most important lessons and best practices on programmes will be highlighted. We provide you with the 20% that will make the 80% difference on your programme and, hopefully, ensure successful completion.

Even though programmes extend over a fairly long period, at some time a programme's objectives will be met or, in unfavourable circumstances, the programme may need to be terminated. Termination and close-out of the programme is also discussed.

Introducing the case study

At this early stage we introduce the case study. This case study is used throughout the book to explain the principles and lessons shared with you. As explained in the preface to the book, a book symbol inscribed Intego Case Study, together with text in light blue, is used whenever work is done on the case study. This is used to separate the details of the case study from the body of text.

For the case study, we make use of a fictional company, Intego Holdings Limited (IHL), which focusses on the generation and distribution of power from coal and natural gas, as well as the direct sale of gas to customers. IHL also holds equity in several joint venture companies along the value chain and buy essential services in from external companies.

In keeping with the proposed layout and for consistency, all further discussion of Intego will be done as described above.

Intego case study

Introduction

Intego has been operating in the power generation arena for the past 50 years and has steadily grown in size and complexity as new plants were commissioned and additional businesses along the overall value chain were acquired. In the past, it has only been necessary to adopt newer environmental standards on new plants, while the environmental permits of the older plants remained in place as per the requirements at the time of their design. New environmental legislation both locally and internationally requires that the company takes an integrated approach towards meeting recent, stricter air pollution prevention requirements.

It has become clear that unless the company can provide a strategic plan and adhere to it, the company's environmental 'license-to-operate' could be in jeopardy. It has therefore been necessary to determine an overall baseline of all key air pollutants and to develop a strategy to keep these emissions in line with the anticipated regulatory changes. It is anticipated that the reworked regulations will be introduced in a stepwise manner up to 2025, requiring more prudent emission control. It has been agreed that from the time of promulgation of the new standards a five year grace period will be allowed to design and install the necessary plant changes to enable those affected to meet the new requirements.

Company structure and value chain

The company structure and value chain for Intego is shown in Figure 3.4. Intego Holdings Limited (IHL) is the holding company directly controlling a mining company (MiningCo) and a power generation company (PowerCo). MiningCo is responsible for all coal mining and gas field operations as well as product classification and/or conditioning. Once the product is ready for transfer, Logistics Pro, an independent service provider, is responsible to truck coal product to PowerCo for the production of electric power. PowerCo comprises a milling facility, boiler station and turbines which produce power. The natural gas from the gas field operations is transferred to customers, one of which being the Intego owned gas-fired power stations. Power produced is transferred via GridCo (JV) to the final utility company, UtilityCo, for sale to end-users.

The key players in the value chain are:

*   **Intego Holdings Limited (IHL):** The overarching holding company responsible for the overall value chain;

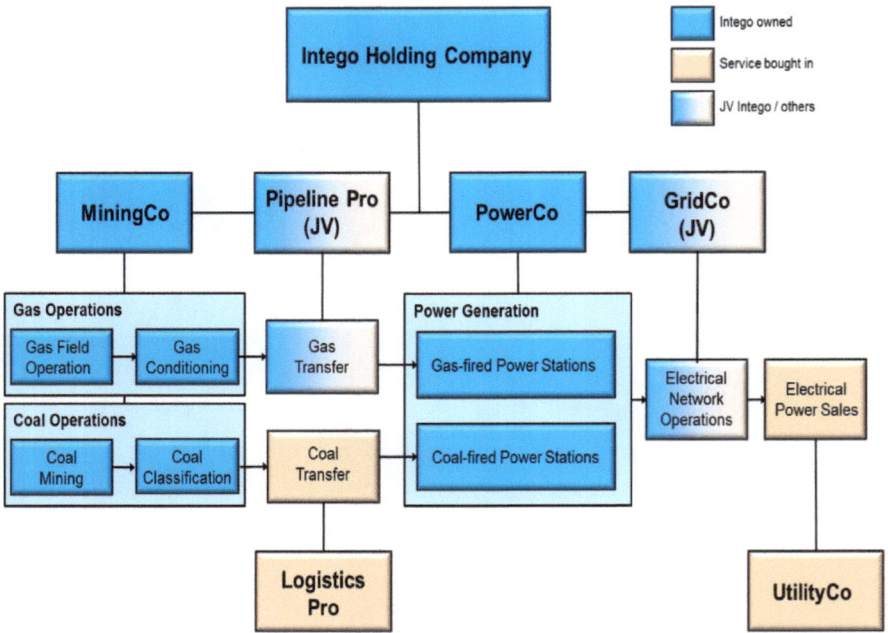

Figure 3.4: Company structure and value chain for Intego

- **MiningCo:** MiningCo is a fully owned subsidiary of IHL and consists of two parts, namely gas operations and coal operations. Gas operations comprise the gas field operations and gas conditioning facilities and the coal operations comprises the mining activity and coal classification;

- **PowerCo:** PowerCo is a fully owned subsidiary of IHL and comprises the power generation facilities of Intego, be it gas-fired or coal-fired power stations;

- **Pipeline Pro:** This is a joint venture company of Intego and external parties for the maintenance and operation of a network of pipelines for the transfer of gas to consumers. A major gas consumer is Intego's gas-fired power stations;

- **GridCo:** This is a joint venture company of Intego and external parties for the maintenance and operation of an electrical distribution network;

- **Logistics Pro:** This is a non-affiliated service provider of Intego charged with the transport of the classified coal feedstock to the coal-fired power stations, and;

- **UtilityCo:** This is a non-affiliated service provider of Intego charged with the sale of electrical power to consumers.

Vision

Intego will continue to supply its customers and maintain its operations in such a way that it will remain within the changing legislative and regulatory environment. Emission of priority air pollutants will be 30% less than the 2010 baseline by 2025 in absolute terms and will conform to interim constraints as regulated. Changes or improvements can be made in any part of the business value chain as deemed necessary to meet the strategic objective.

Boundary conditions within which the team must operate

The company will remain focused on the production of power from natural gas and coal. Other new energy alternatives, for example nuclear power generation, will not form part of this programme. It is essential to ensure that the key cash generating entities will remain sustainable for the next 20 years.

Whenever the Intego case study exceeds one page in length and overflows onto the next page, a solid line, as above, is used to indicate the end of the case study.

Reference is made to templates and pro-forma documents used for the case study at several places in the book. These documents are available to download from our website: www.otctoolkits.com.

The worked case study covers most of the templates and pro-forma documents in detail. However, we reserve the right to continually update and improve the templates and documents on the website.

Concluding remarks

In this chapter we've presented our programme management model and introduced the Intego case study. The eleven elements of the model are discussed in future chapters and the case study is used throughout to illustrate the application of the principles involved.

Part 2, which follows, deals with the six sequential programme management steps in our model, depicted as the six-bladed propeller. The programme lifecycle essentials, depicted as a series of concentric rings in the model, are discussed in Part 3 of the book.

PART 2
Six Sequential Steps

This page intentionally left blank

Chapter 4:
Programme Initiation and Shaping

"All things are created twice; first mentally; then physically. The key to creativity is to begin with the end in mind, with a vision and a blue print of the desired result." - Stephen Covey

Introduction

According to Merrow (2011), shaping a business opportunity means to assess and configure a megaproject (i.e. programme) in such a way that it is profitable for the stakeholder-investors and provides a reasonably stable platform from which to manage the programme. Programme initiation and shaping is thus the logical starting point for a programme; refer Figure 4.1 with shaping highlighted.

Shaping is the first step in the process. Shaping is a core step without which it is not possible to develop and execute a project or programme that will meet a defined business goal. Merrow (2011) further explains that opportunity shaping is a business-led process by which sponsors:

- Evaluate the key attributes of a potential project;

- Develop and gather information that is needed for key decisions, and;

- Allocate the value of the project to the various stakeholders to make the programme environment stable enough for successful execution.

As introduced in Chapter 3, the programme lifecycle as depicted in Figure 4.1 is a continuous process, starting with the initial shaping of the opportunity.

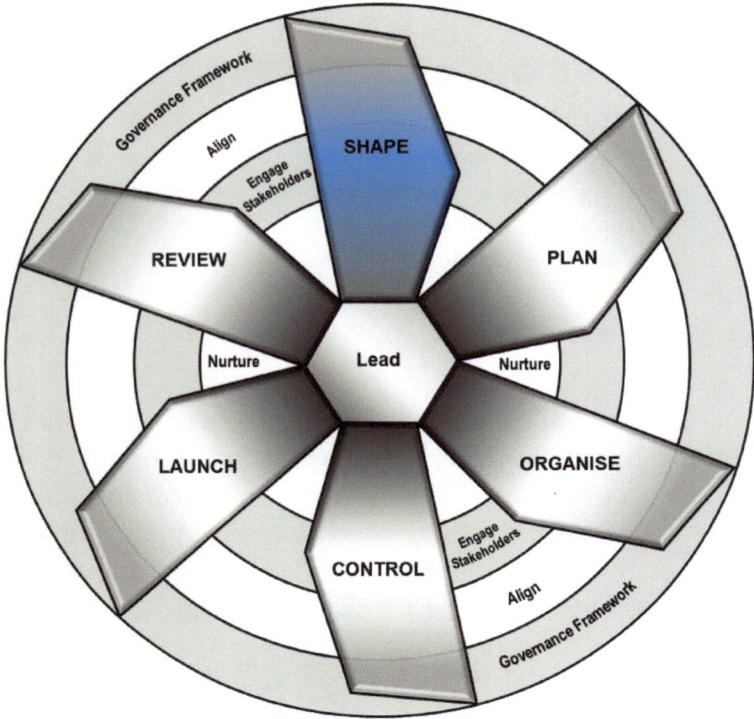

Figure 4.1: Programme management model with 'Shaping' highlighted

Shaping a programme involves two major steps, namely:

- Development of the programme charter, and;

- Development of the programme construct.

The deliverable at the end of each of the steps is a formal document. It is critical that during this shaping process all stakeholders, but especially internal stakeholders from business and project execution, become aligned as to the ultimate objectives and end-goals of the programme.

Developing a programme charter

Introduction

All projects and programmes start off with either an opportunity or problem statement which is translated into an overall business objective and a definition of victory. In order to ensure that a business remains viable, a business strategy is typically developed covering specific objectives like revenue growth, profitability, new product development, environmental and other requirements. This strategy has to be converted into a viable business development plan in order to ensure that the strategy is transformed into an executable plan with tangible deliverables and timelines. In order to communicate these strategies effectively, each has to be developed into a specific sub-set of measurable business objectives. These business objectives must then be translated into more specific and concrete programme objectives covering aspects like programme scope, cost, timeline and others.

The shaping process of a programme starts by getting alignment on the overall business needs/objectives which are captured in a programme charter, followed by an outline of the potential scope, cost, time-line and business impact as seen at the outset of the programme. Upon formulation of the programme shape, formal approval is required from the relevant approval body. It is important to inform the decision makers of the proposed way ahead and more importantly to get their buy-in. The next step will then be a more detailed planning phase with final programme approval.

This chapter deals with the very important first step in the shaping process: developing and agreeing the programme charter that will serve as the guiding document for further development of the programme. After the theoretical discussion of developing a charter, the process will be illustrated through our Intego case study. Shaping a programme is significantly different from that of a single project in that it requires the creation of a framework in which sequential commitments can be made and options exercised as you move forward.

A programme charter needs to be set at the start of the shaping process and although the business objectives should remain firm, the details normally evolve in the first year or two from initiation of the programme. The charter firms up as the team gains more certainty on exactly what will have to be done to achieve the overall business objectives. We need to remind ourselves that a programme typically runs for several years and that it is fraught with uncertainty.

Hence, a programme requires the development of a clear framework plan and tracking of progress in order to ensure that one remains on course towards the end objectives at all times. If issues outside the control of the programme manager changes, it may even be necessary to stop, accelerate or even develop new projects in order to still meet the overall business intent. As such the programme manager and business development team must work in close co-operation at all times and the programme manager needs to be flexible but also ensure that the individual projects already approved are not derailed. Key to a successful programme is the ability to protect the projects during execution but still to be able to devise ways in which changes in business needs and assumptions can be accommodated.

Purpose of the programme charter

In order to align key participants the programme director and the business development leader will facilitate the development of a comprehensive programme definition or charter. The purpose of the programme charter is to document:

- Reasons for undertaking the programme;

- Objectives and constraints of the programme;

- Assumptions underlying the objectives;

- Killer concerns and risks that could derail the programme;

- Directions concerning the proposed solution;

- Identities of the main stakeholders, and;

- Boundaries within which the programme will operate.

In contrast to a specific well-defined project charter, a programme charter tends to outline the overall long-term objectives, boundary conditions and timeline and is developed in more and more detail as specific sub-projects are identified and scoped. This development continues up to the point where the total programme that will meet the business objectives is defined and all projects have been initiated. Certain sub-projects need to be started sooner in order to meet the overall business objectives, while other areas and projects are still being defined. The charter aids in communicating to all stakeholders the development status of the programme and creates a set of common expectations about what will happen when. It helps prevent shortcutting of critical work that must be accomplished to ensure a successful programme. The charter will need updating as key milestones of the programme are being reached and requires continuous review of assumptions, constraints and killer risks to ensure that the programme and its respective sub-projects are still viable. Assumptions outside the control of the programme team need specific emphasis. Should a change in direction be required for the programme, it can be done pro-actively.

During this process the shape of the programme will become clearer and individual projects will emerge, each with its specific project charter, which will then be delegated to specific project teams to execute.

A successful charter is able to clearly convey the message toward an effective change. It is essential to address three key aspects, namely:

- The present situation and the reason for being dissatisfied, without which no reason for change can be visualised;

- A clear vision of where the owner would like to be is required to continuously focus the team and prevent losing momentum, and;

- An outline process of how to get to the desired state to prevent frustration and anxiety as the programme unfolds.

Typical contents of a programme charter

This section will illustrate what a pro-forma programme charter looks like and what the typical requirements are. This is followed by an actual programme charter as developed for the Intego case study.

The proposed programme charter template discussed below consists of four sections, namely:

- **Business Overview section:** The first section describes what the business requires the programme to achieve;

- **Programme Execution Approach:** The second section outlines at a high level the key aspect of how the programme should be set up to achieve these goals;

- **Approvals and signoff:** The third section contains the required approvals and signoff for embarking on the programme, and;

- **Appendices:** The last section contains all appendices, including the programme terminology, references, roadmap, budget and risk register. It is useful to also list all assumptions regarding the programme in this section.

While the business overview section will remain essentially static throughout the life of the programme, the programme execution approach section will be enhanced and further developed by the programme owner team as the programme definition improves. Depending on the nature of your programme, additional sections may be required, but we'll restrict ourselves to examining these components.

Pro-forma documentation will be depicted in italics in the book, as used below.

Pro-forma Programme Charter

1. *Business Overview/Mandate*

1.1 *Project Identification*

Give the formal name of the programme, as well as any other terms that might be used to identify the programme and the primary groups that will be involved with it. Using consistent terminology will enable effective communication for all parties involved with the programme, such as the programme team, stakeholders, and end-users.

1.2 *Background*

Give any background information that will help explain the need for the programme and how it came to be. Make sure that the issues pertaining to the present state and reasons for change are clear.

1.3 *Overall vision*

The vision for the programme should explain exactly what the desired end state is. If possible, describe the vision in such a way that it can be used to measure successful completion. A useful way to think about a vision is to imagine what you would like to read in a newspaper brief (i.e. at most one paragraph summaries) about the programme after completion.

A vision should be inspiring for the programme team and the business itself, providing the necessary excitement, inspiration and impetus to drive the programme through to successful completion. It should be used throughout the programme lifecycle during communication and team building sessions.

1.4   *Business objective(s) and key characteristics thereof*

Using appropriate business terminology, state all sub-objectives (in support of the vision). Be sure to include the programme's intended purpose in language that is both concise and explanatory. Explain how this programme can support achievement of the corporate strategic intent and business unit strategic objectives.

Business objectives should be:

* *Specific – target a specific area for improvement;*

* *Measurable – quantify or at least suggest an indicator of progress;*

* *Assignable – specify who will do it;*

* *Realistic – state what results can realistically be achieved, given available resources, and;*

* *Time-related – specify when the result(s) can be expected.*

In other words, and using the first letter of each of the bullet points above, business objectives should be SMART.

1.5   *Programme Scope*

Programme scope should include detail on the geographical areas covered, departments involved, products to be produced or changed, functionality required and market segments to be affected.

The high level scope of the programme should be clearly defined so that all parties involved are aware of exactly what the programme includes and what it doesn't.   It is also essential to indicate who is responsible for the different parts of the scope, especially where there are joint venture partners, contractors and intellectual property issues at hand.

1.6   *Boundary conditions and constraints within which the team must operate*

List the specific boundary conditions and constraints that must be taken into account.   Boundary conditions specify those limits which cannot be exceeded.   These could for example include issues like maximum permissible emission loads, infrastructure and utility limitations, employee number restrictions, licensing issues and timing of certain parts of the programme.

1.7   *Major assumptions underlying the vision and objectives*

Determine whether there are any specific assumptions made during the initial development of the programme charter's vision and objectives. These assumptions must be made transparent and the team needs to keep track and verify the validity of these assumptions. Any change in the assumptions could have a major impact on the feasibility of the programme. Assumptions are inherent risks to the programme that must be managed carefully and validated before final approval of specific sub-projects if possible.

1.8 Potential influences

Any initiatives or projects within the organisation and the immediate community that may influence or affect this programme should be listed. The team will need to take cognisance of these and monitor the impacts carefully in terms of dependencies, overlaps and conflicts.

1.9 Killer concerns or risks

Killer risks and concerns surrounding the programme should be clearly listed and prioritised. These may in some cases be linked to assumptions made regarding the programme or the technologies used. The team should be clear on what to pro-actively manage to ensure programme success.

1.10 Exit indicators/Abandonment criteria

Specify what criteria could lead to the programme or a specific project's termination. For example, a project's return on investment being below the target hurdle rate or unacceptable safety or environmental risks associated with the programme.

The team needs to develop an exit strategy should the exit criteria actually materialise. If not, the programme or sub-project could continue beyond the point where it is clear that investment should stop, even when it is clear to decision makers that the project will not meet the original objectives set (the so-called 'Abilene Paradox'). A common phrase relating to the Abilene paradox is a desire to not 'rock-the-boat'.

1.11 Key stakeholders to engage

Key stakeholders (also known as interested and affected parties) that need to be involved and informed about the programme's progress should be listed. These include key internal and external stakeholders that have a major stake or interest in the programme and could have a major influence on the programme. Influence could be in terms of funding, resources, impact and regulatory approvals.

1.12 Community relations and corporate social investment

Once the key stakeholders have been identified and as part of the stakeholder engagement strategy, it will be necessary to engage with, and invest in, local communities in order to build relationships and to conform to legislation in this regard. Investment may include community projects (clinics, schools and hospitals), job creation and or small enterprise development. Both the King III Code (IoD, 2009a) and the Sarbanes-Oxley Act (USA,2002) require companies to display their commitment to social responsibility from a corporate governance point of view.

1.13 Legislative and governmental SH&E requirements

It is vital to highlight in-country legislated safety, health and environmental (SH&E) requirements relevant to the industry in question. Where no in-country environmental standards exist, the typical approach is to use the latest environmental, health and safety standards of the International Finance Corporation (IFC). Make it clear to the team what needs to be conformed with.

1.14   *Sponsorship*

Describe sponsorship of the programme. In contrast to a single project, a programme will often be guided by an overall executive sponsor, while the sub-projects are sponsored and owned by different business unit owners or individuals.

Provide a list of names identifying the major parties involved in the programme, such as programme sponsors, and eventual programme owners. In addition, be sure to identify the role of each individual listed so that there is no confusion concerning accountabilities at a later stage.

2   Programme Execution Approach

A charter document should also include a section on the programme execution approach to be followed, providing a high-level overview of how and when the programme will be developed and implemented. There are several standard components that normally comprise the programme execution approach section.

2.1   *Programme management principles*

Explain which of your organisation's programme management methodologies will be employed for this program. If you plan to deviate from the standard programme procedures and methodologies of the company, explain that as well.

The programme control component explains the tools that will be used to assist the programme manager in tracking progress. It also serves as a communication device for communicating the programme's progress to the programme team, executive sponsor, and stakeholders.

2.2 High-level roadmap

Highlight the key deliverables, milestones and expected timeline from an overall integrated programme perspective. Define the expectations in terms of the overall programme timeline as well as the timelines of specific sub-projects or initiatives. Note any specific milestones and decision points that the programme team must take cognisance of. It is suggested that a more detailed programme roadmap is provided as an addendum to the charter. This can then be reviewed and updated regularly as new information becomes available, with the necessary revision control procedure implemented.

2.3 Governance

Explain the governance structure for the programme and how it links to the organisation's governance. Make sure it is clear how the programme direction and requirements will be set and controlled to ensure an overall optimised approach. Explain how each sub-project will be managed and controlled. This should include a high level governance structure indicating which governance bodies or individuals approve what.

Include a summary of key parties that will be directly involved in the programme and what their roles will be.

2.4 Proposed budget

Generally a programme will require an annual budget to cover the overall management of the programme and the running of the programme office. These overhead expenses are generally reallocated to the individual sub-projects on an agreed basis.

Secondly each project as part of the overall programme requires a specific budget and capital expenditure to cover the project specific needs.

Highlight any issues or assumptions regarding the budget for the development of the programme running costs as well as overall capital cost, if applicable. It is suggested that a more detailed programme budget be provided as an addendum to the charter.

2.5 Programme management resources

Specify what other key resources, in addition to human resources, the programme management team will need for satisfactory co-ordination and management of the programme. Provide a view of key resources that might be over utilised or may be critical during the programme's lifecycle.

2.6 Risk management

Discuss the approach that will be followed for managing risk for the programme. In contrast to a single project, a programme will have to consider overall programme risks on an integrated basis, individual project risks and collation of common risks across projects and escalation to the programme level. How will project level risks be identified and managed vs. programme level risks? How will risk be reviewed to ensure that common risks across projects are escalated to the correct level? How will risk management be reported on?

It is suggested that a more detailed risk register be provided as an addendum to the programme charter.

2.7 Stakeholder management

Stakeholders are engaged differently based on their level of interest and concern in the programme and the influence or power they may have on the programme. Depending on these criteria, the engagement strategy for each of the stakeholders or sets of stakeholders will differ. Describe the methodology to be followed in developing a stakeholder management plan and the management thereof.

2.8 Business architecture

Megaprojects executed as a programme will normally significantly affect the overall business. Describe any issues in terms of a blueprint for the various aspects of business change, e.g. people, organisation structure, technology and processes that may be of significance.

3 Approval Section

The charter will be recommended by the members of the programme steering committee and approved by the chairman of the steering committee. The steering committee chairman is typically the executive sponsor. The approval section of a programme charter may be the simplest section to put together, but it is one of the most critical in terms of the programme's eventual success.

List all of the names and roles of the major stakeholders along with their signatures, indicating that each of these individuals is satisfied with the details included in the charter.

In addition, if the programme will require resources from other departments or groups, a representative from each of these divisions should be listed in the approval section as well. The signatures of these individuals will signify that they accept their own responsibilities for successful completion of the programme and agree to provide the support as specified in the programme charter.

The programme leadership team will also sign the charter as proof of their acceptance of the targets and requirements outlined.

4   Appendices

4.1   References

List any related documents or other resources that could be helpful in understanding various aspects of the programme, such as the scope, business need, or legislative requirements.

4.2   Terminology

Provide a basic, but complete, glossary of terms that defines special terms related to the programme. If there are any key terms, phrases, or acronyms that might prove to be confusing or new to anyone related to the programme, be sure to include them in this section.

4.3   Project Roadmap

Provide a copy of the high level programme roadmap with key deliverables and milestones.

4.4   Project Budget

Provide a copy of the high level programme budget for the programme running costs and capital expenditure.

4.5   Risks

Provide a copy of the integrated programme risk register with the high priority risks identified.

Uses of a programme charter

A programme charter developed according to the pro-forma template described above will have three main uses, namely:

*   To authorize the programme and project teams to kick-off the work required as per the mandate given to them in the charter;

*   To serve as the primary programme marketing document. Stakeholders, primarily programme team members, have a 1-to-2 page summary to distribute, present, and keep handy for fending off other projects or operations competing for resources, and;

*   To serve as a focus point throughout the programme to ensure tight scope management against the original approved programme charter. It almost becomes a business baseline document, even though many more detailed documents are required to support the information and assumptions contained in the charter.

Although the main effort towards programme definition and shaping will occur at the start of the programme, the business context may evolve over time and circumstances may change. Those parts of the programme definition that handles either the overall business solution or how it will be achieved should be viewed as an evolving model. This should be managed actively during the programme in order to achieve the optimum overall benefit for the organisation.

Please note that Chapter 13 on alignment cannot be seen separately from this chapter. Every aspect covered in this chapter is an element on which alignment should be gained between all relevant stakeholders.

Now that we have seen what the typical requirements for a programme charter are and what a charter is used for, it is time to return to our case study. The management team of Intego decided that the strategy regarding environmental compliance of its operations as discussed in Chapter 3 will be developed and implemented as a programme. This will be done in an integrated approach for all companies in the value chain. The programme will cover the total timeline up to the envisaged end-point of 2025. A 15-year programme is thus envisaged, covering multiple projects at multiple sites where the company is involved.

A programme director was appointed to join the Intego business development team to turn the high level business strategy into a clear statement of business intent and business objectives and to translate the business needs into clear programme objectives. The first step was to disseminate the business strategy and business objectives. This was done in an interactive framing workshop (refer Chapter 13 on alignment) from which the programme charter was finalised and issued.

The complete IECP programme charter follows.

Intego Environmental Compliance Programme (IECP)

Programme Charter:

1   Business Overview/Mandate

1.1   Project Identification

The programme is named the Intego Environmental Compliance Programme (IECP). All communication and documentation must refer to either the full name or the abbreviation.

1.2   Background

Intego has been operating in the power generation arena for the past 50 years and has steadily grown in size and complexity as new plants were commissioned and additional businesses along the overall value chain have been acquired. In the past it was only necessary to adopt newer environmental standards on new plants while the older plant's permits remained as per the requirements at the time of their design. The new environmental legislation both locally and internationally requires that the company takes an integrated approach toward meeting newly regulated and stricter environmental requirements.

It has become clear that unless the company can provide a strategic plan and adhere to it, the company's environmental 'license-to-operate' could be in jeopardy. It has therefore been necessary to determine an overall baseline of all key air pollutants and to develop a strategy to keep these emissions in line with the anticipated regulatory changes. It is anticipated that the reworked regulations will be introduced in a stepwise manner up to 2025, requiring more prudent emission control. It has been agreed that from the time of promulgation of the new standards a five year grace period will be allowed to design and install the necessary plant changes to enable those businesses affected to meet the new requirements.

1.3 Overall vision

Intego will focus on the priority air pollutants associated with the mining and power generation sectors, namely oxides of sulphur (SOx), oxides of nitrogen (NOx) and particulate material, particularly the smaller PM-10 particles that are deemed to be responsible for adverse health effects because of their ability to reach the lower regions of the respiratory tract. The PM-10 standard refers to particles with a diameter of 10 micrometers or less.

Intego will maintain its operations in such a way that environmental emissions from their facilities will remain within the changing legislative and regulatory standards. Emissions of priority pollutants will be 30% less than the 2010 baseline in absolute terms by 2025 and will conform to interim constraints and limits as regulated. Changes or improvements can be made in any part of the Intego business value chain in order to meet the stricter environmental standards.

1.4 Business objective(s) and key characteristics thereof

The objective is to reduce SOx, NOx and particulate matter (PM-10) emissions by 10% in 2015 (Phase 1), 20% in 2020 (Phase 2) and 30% in 2025 (Phase 3) as compared to the 2010 baseline.

Total additional cost (operating and capital) is to be optimised and calculated on an annualised basis (capital written off plus annual operating cost) and should not cause an increase of more than 15% per production unit. Total capital required for the changes is not to be considered a constraint as long as the specific investments can meet the company's published hurdle rate.

Overall output for the 'in scope' areas in total may not be lower than the agreed published annual, monthly and weekly budgets.

Safety is a key priority and the programme is to proceed on the basis that all accidents can be prevented and the target for the total programme is zero fatalities and a recordable accident case rate of less than 0,2.

1.5 Scope

The programme will include all operations directly or partially owned via a joint venture company by Intego, namely:

- MiningCo;

- PowerCo;

- Pipeline Pro, and;

- GridCo.

Other companies in the value chain operating independently are excluded, namely:

- Logistics Pro, and;

- UtilityCo.

The scope of the programme is to include the priority pollutants as per the regulations:

- Oxides of sulphur (SOx);

- Oxides of nitrogen (NOx), and;

- PM-10 particulate matter (commonly known as ROx).

Only operations within the boundaries of the Republic of South Africa will be considered.

The boundary for measurement of emissions is limited to emissions that are produced from the direct manufacturing units and any co-incidental emissions are not included. Thus the emissions generated by, for example, vehicles owned by the company are excluded.

1.6   Boundary conditions within which the team must operate

The company will remain focused on the production of power from natural gas and coal. Other new energy alternatives, such as nuclear, solar and wind power generation will not form part of this programme. It is essential to ensure that the key cash generating entities will remain productive for the next 20 years. Advanced technologies for coal to power generation are to be considered as well as innovative or new less proven technologies to reduce the environmental footprint. As the programme extends over a number of years, technologies currently still in the research and development stage can also be considered in the technology roadmap.

It is expected that competent resources to operate these complex units, will become more and more difficult to obtain. Therefore advanced control, artificial intelligence and optimisation systems may be required to reduce reliance on large numbers of highly skilled personnel. Construction labour requirements are to be identified and specific training programmes implemented to ensure sufficiently skilled personnel will be available without the need for importing expatriate construction workers.

Staff establishment is of specific concern and the programme must not result in the total permanent labour force to increase above the approved 2010 numbers.

Although this programme's focus is on reducing emissions to atmosphere, it could lead to additional aqueous effluent being generated. The absolute mass load of regulated substances being released at each facility cannot be increased above the present load.

Community and labour relations and local economic development are critical to the success of the company. Well-managed community programmes and interaction is expected with zero disputes or incidents as the target. A pro-active approach in dealing with local communities is a requirement.

1.7 Major assumptions underlying the vision and objectives

It is assumed that the reduction in emissions will proceed as indicated in the outline presented by the Department of Environmental affairs. The programme planning should proceed on this basis but none of the proposals have yet been published as final legislation. Lobbyists or government objectives could change the actual final environmental requirements. The programme team must actively track the legislative developments and inform the stakeholders of changes. It will be the responsibility of the programme team to generate proposals on how to adapt the programme to suit the changes.

A major assumption is that government will opt for regulations lower than the business objective set in this charter, namely reduction of SOx, NOx and particulate matter by 10% in 2015,

20% in 2020 and 30% in 2025. A strategic decision was taken to take a progressive stance to also accommodate possible future changes in legislation.

1.8 Potential influences

Business improvement projects and practices at the various operating locations may result in changes to the baseline emissions from these sites. As the programme extends over a number of years, potential influences could include:

- keeping track of developments at the various sites;
- ensuring alignment with the overall programme objectives, and;
- the impact of changes in one component of the value chain on other components.

1.9 Killer concerns or risks

Changes in the proposed environmental legislation could result in major adaptions and directional changes in the programme objectives.

The financial cost associated with implementing stricter than anticipated changes in the environmental standards may not be viable.

1.10 Exit indicators/Abandonment criteria

Should government relax their requirements extensively, a programme approach could be avoided and minor adjustments made at the different sites to comply with the emission standards.

Financial feasibility is a major concern. Creative ways needs to be found to deliver on the legal requirements and still have it make business sense. It may imply that some units close down if it makes business sense.

1.11 Key stakeholders to engage

External stakeholders include:

- Shareholders;
- National Government: Department of Manpower, Department of Environmental Affairs, Department of Public Health;
- Regional Government: environmental affairs, manpower and public health;
- Local Government: environmental affairs, manpower and public health;
- Relevant non-governmental organisations and environmental pressure groups;
- Organised labour;
- Local communities;
- Medical practitioners;
- Media;
- Air quality specialists;
- Competitors;
- Technology suppliers;
- Feedstock suppliers;
- Customers, and;
- Any other interested and affected parties.

Internal stakeholders include:

- Intego CEO;

- Financial director;

- Programme sponsor;

- Public relations group;

- Corporate social investment group;

- Managing directors of MiningCo, PowerCo, Pipeline Pro and GridCo;

- All Intego personnel, especially mining and operating personnel;

- Safety, health and environmental specialists, and;

- Programme team.

Other companies in the value-chain, operating independently from Intego, are excluded, but should be informed of the changes; these are Logistics Pro and UtilityCo.

1.12 Community relations and corporate social investments

A pro-active approach will be followed to engage the local communities via the formal and informal community engagement structures that exist. Bi-monthly feedback will be provided to local communities on the progress of the programme and on the possibility of creating jobs for local labour, and local economic development. A formalised tender process will be followed in assigning contracts to all contractors.

Environmental impact assessments for any proposed changes will be done by external impact assessors according to the published guidelines.

1.13 Legislative and governmental requirements

It is anticipated that the environmental regulations and stricter emission standards regulations will be introduced stepwise up to 2025, requiring more stringent control of emissions to atmosphere. Even though the legislative requirements have not been promulgated yet, all indications are that should Intego achieve their business objectives as stated above, namely reduction of SOx, NOx and PM-10 particulate matter by 10% in 2015, 20% in 2020 and 30% in 2025, Intego should be well within the revised environmental requirements. These are tracked continually to ensure compliance and adjustment of the programme, if required. It is assumed that Logistics Pro and UtilityCo will not be affected by the changes in the rest of the value chain.

1.14 Sponsorship

The programme will be sponsored by the company CEO and he will chair the quarterly overall programme integration steering committee. The CEO will remain the decision executive for the shaping of the overall programme supported by the managing directors of the different business units. As the different projects will have different needs and be executed on different timelines, each project that obtains final investment approval will be sponsored by the managing director of the relevant business unit within the guidelines laid out for the overall programme.

2 Programme Execution Approach

2.1 Programme management principles

The company standard project execution model will be used as the overall guideline for the programme as well as for the individual projects with the agreed phases and approval/hold point adhered to. At a programme level, it must be considered that the overall implementation of the programme requires various phased approvals of the individual projects. The programme director must propose an overall governance methodology at the first steering committee to indicate how he will ensure and report back on the overall health of the programme.

The programme will be controlled using standard company cost and schedule control procedures. Once an individual project is approved it will be executed and controlled against the approved project baseline. For overall control of the programme, sub-project costs and schedules will be rolled up to the overall programme level. Projects will be reporting against monthly targets using the standard company reporting format.

2.2   High level roadmap

The programme must meet the following overall timeline in terms of key deliverables and milestones measured in months from sign off date of this programme charter:

- Technology calendar: 2 months;

- Legislative calendar: 2 months;

- Overall framework plan linking to above 2 elements: 4 months;

- Indicative front-end loading study 1 (FEL1 or pre-feasibility) business evaluation with costs, economics and impacts: 9 months;

- Sign-off of first phase investment plan and framework: 12 months, and;

- Final investment decision for each step of programme: 3 years ahead of legislative date.

The overall timeline is outlined in Figure 4.2 below. The overall methodology is to develop a broad outline of the total programme, the technology options to be considered, very rough capital and operating costs and overall business impact as the programme's overall shaping study.

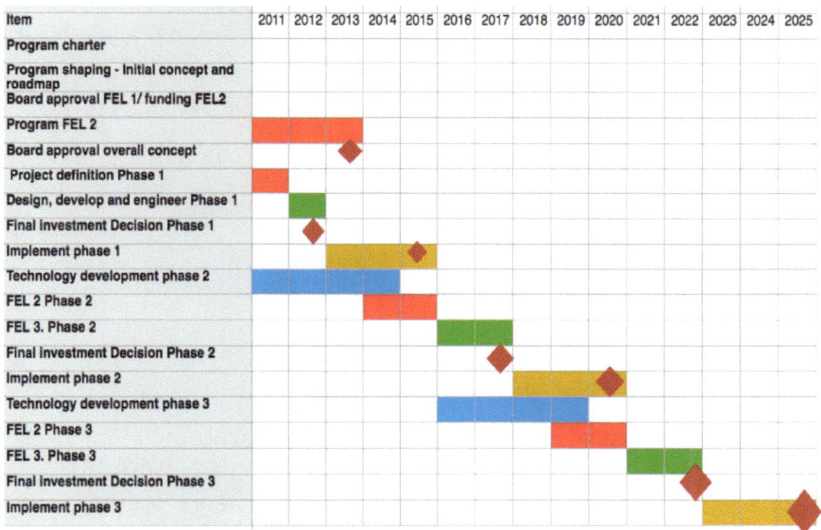

Item	2011	2012	2013	2014	2015	2016	2017	2018	2019	2020	2021	2022	2023	2024	2025
Program charter															
Program shaping - Initial concept and roadmap															
Board approval FEL 1/ funding FEL2															
Program FEL 2															
Board approval overall concept															
Project definition Phase 1															
Design, develop and engineer Phase 1															
Final Investment Decision Phase 1															
Implement phase 1															
Technology development phase 2															
FEL 2 Phase 2															
FEL 3. Phase 2															
Final Investment Decision Phase 2															
Implement phase 2															
Technology development phase 3															
FEL 2 Phase 3															
FEL 3. Phase 3															
Final Investment Decision Phase 3															
Implement phase 3															

Figure 4.2: Timeline for the IECP

This construct will be presented to the holding company as well as the sub-company boards for approval and release of funding for programme planning work. The intent of this approval is to ensure overall alignment as to the proposed direction and potential impact of the programme. Because of the short time to implement phase 1 (2015 end-point), the project definition work for phase 1 must start immediately after board approval, in parallel to the overall programme planning phase development.

As a result of the schedule required to meet the phase 1 objectives, the final investment decision and release of full funds for the implementation will have to be taken well before the total development of the overall concept has been completed. Phases 2 and 3 can entertain more developmental work and will be approved as indicated.

2.3 Governance

The programme and individual projects comprising the programme will be governed through a set of steering committees.

The overall programme integration steering committee is the point from where the overall co-ordination, communication, direction, timelines, budget and resources are set and monitored. Each sub-project will be steered through a project specific steering committee that must ensure the project is executed within the overall guidelines as set by the integration steering committee and also ensures that the specific project is executed within the project specific mandate.

The roles and accountabilities of the committee members of the programme steering committee are outlined in Table 4.1.

Table 4.1: Composition of the Programme Steering Committee

Representative	Role and Accountability
Chairman (sponsor, decision executive)	Ultimately responsible for the business success of the programme.
Operations/Business owners	Eventual business owner accountable for ensuring an operable and maintainable facility. Depending on the

Representative	Role and Accountability
	scope of the programme, more than one operations and business owner may be involved.
Chief Financial Officer	Accountable for ensuring economic viability and cash flow management for the programme.
Chief Technology Officer	Accountable for ensuring appropriate technology roadmaps, innovation and technology effectiveness.
Owner Programme Director	Accountable for sound programme planning, development and implementation practices.
Ad hoc members	The chairman can co-opt ad hoc members with specialised knowledge as may be required to give input into a particular topic/decision at meetings.
Risk Manager	Accountable for integrated risk management across the programme.

2.4   Proposed budget

Since the work envisaged within this programme centres on future environmental legislative requirements, it is required to ensure that all expenditure (both programme running costs and capital expenditure) are controlled carefully to minimise the actual cost of the programme as far as possible. No specific budget or limits have been envisaged and the project team needs to develop an initial view for consideration by the board.

2.5   Programme resources

The facilities to set up a programme management office will be made available at the Intego head office. The programme team must set up a fully functional programme office for the IECP using an open plan office layout that will be suitable for the long-term nature of the programme.

All necessary tools for project management, as well as engineering and communication tools required need to be identified, approved, acquired and installed. This is essential considering that various sub-projects will be managed and developed from different engineering offices and manufacturers globally. Document management and collaboration technology is a critical part of the tools required.

2.6   Risk management

A key risk to the programme will be the late delivery of specific pollution reduction initiatives with the resultant potential penalties enforced by the authorities or closure of the facilities due to non-compliance. The risk to the successful implementation of the programme needs to be thoroughly developed by the risk manager using the company's specific risk management procedures.

The risk of the project not being financially viable should be managed carefully on all fronts.

2.7   Stakeholder management

The formal corporate structures for engagement of external stakeholders will be followed at all times. For the programme, a stakeholder identification and management plan will be developed to clearly map all key stakeholders in terms of their interest/concern and influence on the programme. This will be supported by a clear communication and engagement structure with clear roles and responsibilities, and flow of information.

The official guidelines for engaging with interested and affected parties, as published by the Department of Environmental Affairs, will be used for the environmental impact assessments.

3   Approval

We, the undersigned, acknowledge the strategic importance of the IECP for the long-term survival and license-to-operate for Intego. We commit ourselves to the successful completion of the IECP according to the charter.

Sponsor: ....
                           Signature   Date

Business manager: ....
                           Signature   Date

Programme Director: ....
                           Signature   Date

Developing a programme construct

Introduction

Once the programme charter has been completed and in-principle agreement obtained, the next important step is to develop an initial view of what the programme will entail from the perspective of the owner team.

Often it is found that the first concept as developed during initial work cannot be fully satisfied once more focused planning is done. Adjustments may be necessary to the programme construct, requiring several iterations, until the programme details and practical implications and the business needs are fully aligned.

Typical programme construct

The programme construct is developed jointly by the business development and project execution professionals on the owner programme team.

Finalising the shaping process now requires the following steps:

• Determine technologies available and possible total programme scope;

• First very high level cost estimate;

• Master schedule or framework plan (this is a very high level schedule), and;

• First round of business option evaluations and agreement on potential options to include in further evaluations.

How to actually go about finalising the above topics will be covered in more detail in the subsequent chapters and the focus in this chapter will be on the necessary outcomes to ensure that the shaping process is properly completed.

At this stage we return directly to the Intego case study to show what a typical programme construct comprises. For purposes of the case study, we specify that Intego currently operates five 3000 MW coal-fired power stations and three 1000 MW gas-fired power stations.

Intego Environmental Compliance Programme (IECP)

Programme Construct

1. Introduction

Table 4.2 shows the percentage of total emissions of the priority air pollutants of interest, namely NOx, SOx and PM-10 particulates, for the entire Intego organisation. The three columns in the table each adds up to 100%. As is to be expected, the older coal-fired power stations (Coal 1 and Coal 2) are the most problematic as far as emissions are concerned. However, it also shows where the greatest impact can be made by an intervention such as the IECP.

Table 4.2: Primary sources of air pollutants

Source	% of total NOx	% of total SOx	% of total PM-10
Coal-fired Power Stations			
Coal 1	25	25	27
Coal 2	20	20	23
Coal 3	10	15	15
Coal 4	10	12	14
Coal 5	10	10	14
Gas-fired Power Stations			
Gas 1	10	5	1
Gas 2	5	5	1

Source	% of total NOx	% of total SOx	% of total PM-10
Gas 3	5	5	5
Other sources			
Mining	2	0	0
Gas production	3	3	0

NOx = Nitrous oxides; SOx = Sulphur oxides & PM-10 = Particulate matter <10 micrometers

2. Technology options

In order to meet the interim standards for these pollutants by early 2015, the following opportunities have been identified.

2.1 For reducing NOx emissions

By converting gas-fired power stations to low NOx burners a reduction of 50% in Gas 1 and 30% on Gas 2 and Gas 3 is possible. This results in a total reduction of 8% of total NOx emissions. Estimated cost for this conversion to low NOx burners is USD 90 million and will take 36 months to complete.

Conversion of Coal 3, Coal 4 and Coal 5 to low NOx burners will result in a reduction of 10% NOx per unit, i.e. a total reduction of only 3% of total NOx emissions. Estimated cost for this conversion is USD 50 million per unit and will take 48 months to complete.

As far as Coal 1 and Coal 2 are concerned, the basic technology employed is too old and not suitable for conversion to low NOx burners. Emissions from these two power stations will be addressed in phases 2 and 3 of the programme.

2.2 For reducing SOx emissions

The installation of a sulphur removal unit in the natural gas supply to Gas 1, Gas 2 and Gas 3 from the gas production unit will result in a reduction of 90% of SOx emissions from the gas-fired power stations. This results in a reduction of 13,5% of total SOx emissions. Estimated cost for this conversion is USD 70 million and will take 42 months to complete.

A switch to an alternate low-sulphur coal supply to Coal 2 and Coal 3 power stations will achieve a reduction of 30% SOx on each unit. This results in a reduction of 10,5% of total SOx emissions for both units combined. Duration to expiry of existing coal supply contract is 36 months. The low-sulphur coal is expected to be 10% more expensive than the current feedstock.

The installation of removal facilities for hydrogen sulphide (H_2S) on all coal-fired power stations will return a 20% reduction in SOx emissions for each. Estimated cost for this conversion is USD 100 million for each coal-fired power station, i.e. a total expenditure of USD 500 million, and will take 50 months to complete.

2.3 For reducing PM-10 particulate matter emissions

Short-term solutions include the installation of electrostatic precipitators on any coal-fired power station, thereby reducing PM-10 particulate matter emissions by 75%. Estimated cost for this conversion is USD 70 million per installation and each will take 48 months to complete.

In the medium-term coal-fired power stations can be retrofitted with fluidised bed boilers that will reduce the particulate emissions by at least 70%, with an increased overall efficiency of coal to power of 48% vs. the 35% efficiency of the present older units. Estimated cost of these boilers is USD 30 billion per 3000MW installed capacity. Estimated project duration is 60 months per installation.

In the longer-term, new emerging plasma-arc technology, underground coal gasification with synthetic gas to power or high temperature coal gasification with synthetic gas to power can be considered. Estimated total cost for any of these options will be around USD 120 billion. Overall efficiency is expected to be around 55% and emissions around 75% less than that of the current installations. Expected project duration is 72 months, including development time.

3. Master schedule

The scope of the changes required must cover the business areas as defined in the charter. This is then followed by a framework plan outlining the key projects, their timelines and interdependencies.

The overall master schedule for the development of the programme after initial shaping is shown in Figure 4.3.

The roadmap indicates the timeline and potential capital costs. From this it is clear that the phase 1 development must enter the project definition phase as soon as possible. The subsequent phases and development of the overall programme in the medium-term future needs to be further clarified in terms of the overall route that the company needs to take in order to meet its overall business objectives.

The level of retrofitting vs. total replacement of older plants affects the amount of capital required to a very large extent, but also changes the future operation costs and further growth opportunities significantly. In order to evaluate these options and develop the route ahead, the programme team, staff mobilisation and competencies of the personnel need to be taken into account to develop the overall programme strategy and plan.

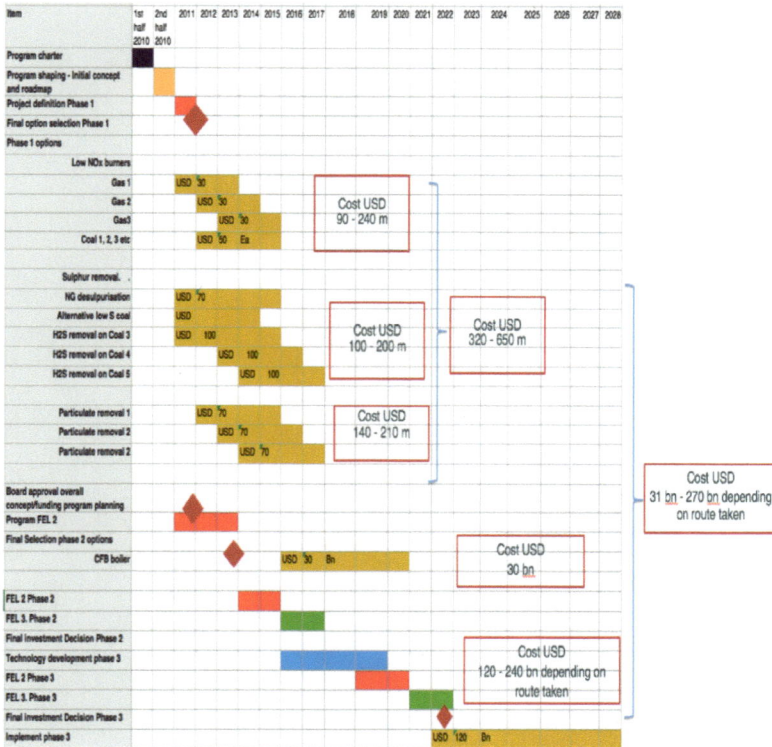

Figure 4.3: IECP Master schedule

The rough indication of potential emission reductions against a timeline are shown in the Figure 4.4, which shows the downward trend for each of the three pollutants from the 2010 baseline; the baseline is indicated as 100% of emission load.

Based on the above roadmap it has been determined that the overall programme team will consist of 50 full-time personnel for a period of 5 years, reducing to 30 full-time personnel thereafter for the next 10 years. The running cost of this programme office for the first 5 years will amount to USD 10 million per year.

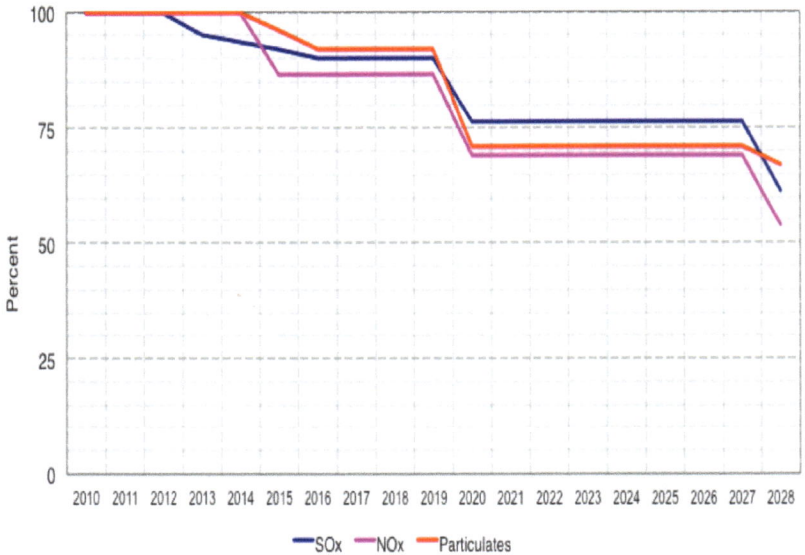

Figure 4.4: Emission profile

The total capital expenditure required is largely dependent upon the overall strategy that is agreed. With the current information, the minimum capital expenditure is approximately USD 1 billion for retrofitting all existing power plants. An intermediate option is estimated at USD 30 billion and an ultimate leading-edge technology option at approximately USD 150 billion.

Table 4.3 gives an indication of the potential cumulative cash flow for the three options in USD millions.

The objective of the programme team will be to further develop the overall programme as per the milestones set out in the roadmap and also oversee the execution of the individual projects as they are kicked off.

Table 4.3: Estimated cumulative cash flows

Year	Cumulative cash flow (in USD millions)				
	2010	2015	2020	2015	2030
Minimalistic approach	0	333	913	1013	1013
Intermediate approach	0	333	30913	31013	31013
Leading-edge Technologies	0	333	30913	99013	151013

Concluding the shaping process

With the programme charter and programme construct completed, this now concludes the first round of shaping the programme.

The business charter has been developed and an initial view of what the programme construct could look like has been developed. There is an early appreciation of the cost and schedule impacts. If required, the charter and construct, specifically the programme execution approach, can be revised at this stage to take into account any additional information that has come to light.

The next step in the shaping process is to get alignment amongst all key stakeholders. Once this has been done, the next revision of the programme charter and execution approach can be issued and signed off. Obtaining alignment requires special skills in facilitation and engagement with stakeholders to ensure expectations are clearly understood, that the programme will address the expectations and concerns and that these will be managed effectively. These aspects are covered in the Chapter 13 on the alignment process.

Chapter 5 on planning will also further elaborate on the actual feasibility of the programme charter and construct. It will show how the gaps between the business wishes and reality, that may become apparent as planning proceeds, can be bridged.

The shaping process ends when all stakeholders: investors, regulators, NGOs, local bodies and governments agree to the programme proceeding as planned and outlined.

The consolidated programme charter and templates can be accessed on the Owner Team Consultation toolkits website: www.otctoolkits.com

Concluding remarks

Programme shaping requires continuous interaction with all stakeholders, including those not so obvious that may lay a claim on project value.

Merrow (2011) states that the intent of the shaping process is to arrive at a point where the following three objectives have been accomplished:

- All of the stakeholders or claimants on project value are either content with their value allocation from the project or have been rendered unimportant; that is, they are no longer stakeholders;

- The project environment has been stabilised sufficiently so that the programme manager will not continually be fighting battles with disgruntled stakeholders while trying to execute the programme, and;

- With these things accomplished, there is still sufficient value in the programme for the owner organisation that its board will ratify the programme going forward without qualification.

This basic programme charter and execution approach can be adapted to fit any organisation and any array of projects, including technology-driven projects. The ease and flow of project completion is directly due to the care taken with the preparation of a programme charter. A good programme charter literally keeps everyone involved in the programme focused on the same objectives.

This page intentionally left blank

Chapter 5:
Planning the Programme

"Plans are only good intentions unless they immediately degenerate into hard work". - Peter Drucker

Introduction

In this chapter on programme planning, the focus will be on developing a clearer definition of the scope of the programme and the resultant schedule. Other elements (e.g. setting up the programme organisation and programme management office) normally involved in finalising a complete programme execution plan will be covered in subsequent chapters. As shown in Figure 5.1, the more detailed programme planning follows the initial shaping process and starts as soon as the core programme team has been mobilised.

Before discussing planning in the context of programme management, it is necessary to ensure alignment on some general project management terms and definitions.

A project management plan is defined as:

- A formal, approved document used to guide both project execution and project control. The primary uses of the project plan are to document planning assumptions and decisions, facilitate communication among stakeholders, and document approved scope, cost, and schedule baselines. A project plan may be summarised or detailed (PMI, 2013a), and;

- A statement of how and when a project's objectives are to be achieved, by showing the major products, milestones, activities and resources required on the project (OGC, 2009).

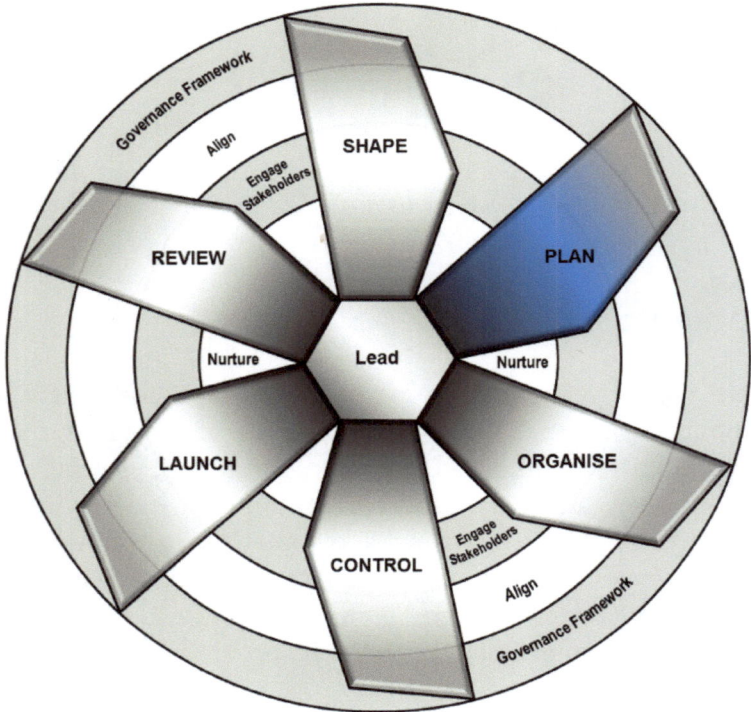

Figure 5.1: Programme management model with 'Planning' highlighted

The project manager creates the project management plan following input from the project team and key stakeholders. The plan should be agreed and approved by at least the project team, the sponsor and the project's key stakeholders.

Project planning is a discipline for stating how to complete a project within a certain timeframe, usually with defined stages, and with designated resources. One view of project planning divides this activity into:

- Setting objectives (these should be measurable);

- Identifying deliverables;

- Planning the schedule, and;

- Making supporting plans.

The project schedule is the tool that communicates what work needs to be performed, which resources of the organisation will perform the work and the timeframes in which that work needs to be performed. The project schedule should reflect all of the work associated with delivering the project on time. Without a full and complete schedule, the project manager will be unable to communicate the complete effort, in terms of cost and resources, necessary to deliver the project. Table 5.1 below expands on the different levels of schedules, use thereof and software tools employed. As the level of the project schedule goes up from level 1 to level 4, so too the level of detail contained in the schedule increases.

Table 5.1: Showing schedule levels and typical tool usage

Level of Detail	Description	Tool Usage
Level 1	Master Schedule/ Framework Plan	Can use MS Excel, MS PowerPoint, Visio, Milestones or other graphics tools to do an overall roadmap, framework plan or master schedule. This high level framework plan or roadmap is typically referred to as a master schedule.
Level 2	Area/Unit Schedules	As a minimum, a formal planning tool (MS Project / Primavera) is used which is set up to easily translate information to the Master Schedule and easily reads information from the level 3 schedules. This is a more detailed breakdown of the Master Schedule and typically follows the project WBS where areas, sub-units or packages are shown. It may also indicate summary discipline level activities.

Level of Detail	Description	Tool Usage
Level 3	Control Level Schedules	The control level schedule is a further breakdown of the level 2 schedule and is where the primary interface between the higher level and lower level schedules occur.
Level 4	Construction Contractor Schedule	These schedules are developed in support of the level 3 schedules and may or may not be included in the level 3 schedule. The level 3 schedule however is developed to easily accommodate the information required from the level 4 schedules. Level 4 schedules are generally developed by construction contractors/discipline leads to detail their daily work in support of the level 3 schedules.

The highest level of the plan is typically referred to as a master schedule. The master schedule is an integral input document for the planner into the overall plan. It is a list of key activities/milestone dates per phase giving an overall, integrated view of the project or programme on one page. The master schedule is compiled and regularly updated from more detailed schedules to depict the latest state of affairs. Master schedules can be used to effectively communicate with executive managers.

The project schedule is the tool that communicates what work needs to be performed, which resources of the organisation will perform the work and the timeframes in which that work needs to be performed. The highest level of the plan is typically referred to as a master schedule.

Programme planning

Scope definition

Now that project planning is understood, let's explore programme planning. Once the scope of the programme has been defined the development of the programme master schedule can commence.

Project teams are often very reluctant to tackle the somewhat tedious process of defining (writing down) the scope of facilities and getting it approved. It is seen as a somewhat unnecessary step since everyone knows what is intended. Neglecting the formal definition of the scope of a programme is a very dangerous and often fatal approach that leads to misunderstanding, rework and serious clashes between stakeholders, project team members and contractors.

A first iteration of the plan and the cost estimate is very useful in creating a perception of the scope and cost implications of any programme, even if it means that in the early phases of a programme one bases those plans and estimates on published or historical timelines and costs. This can help the team set the direction in which the programme team should focus.

The scope definition manual becomes the guidebook for the project management team against which the final deliverables and scope management are measured. It is an invaluable tool and needs to be used daily in explaining, agreeing changes and impact thereof and defending the scope.

Neglecting the formal definition of the scope of a programme is a very dangerous and often fatal approach that leads to misunderstanding, rework and serious clashes between stakeholders, project team members and contractors.

Scope of facilities breakdown structure

The first step in programme planning is developing a programme level Facilities Breakdown Structure (FBS). The FBS should indicate all of the projects and sub-projects making up the programme with a single page definition of each work package. This is illustrated below using the Intego case study.

This work starts with the creation of a breakdown structure that depicts the individual work packages and describes the content of each. As the programme charter and concept is being developed, the scope documentation needs to be created in parallel. A major programme or megaproject is so complex that it is not possible to conceptualise the scope and implications thereof without breaking it down into separate, clearly defined work packages. The packages should be defined at a level of detail from which the first high level estimates and schedules can be derived on 'reasonably' justifiable grounds.

Before a sub-project can be handed over to a project team to execute further, the respective work package must be clearly defined as to the objective, scope and battery limits of the sub-project. The programme management team needs to focus on defining the sub-projects and the battery limits of each sub-project, such that the sum of all the work packages will reflect the total scope of the programme. Nothing should be missed and all interfaces must be clearly defined. At programme level, a work package could still be a very large project. It is not the job of the programme team to break down the sub-project into its more detailed work packages, but that of the individual project teams.

Developing the scope of facilities at programme level must however be aligned with the contracting strategy and the way in which the work packages will be developed.

Intego Environmental Compliance Programme (IECP)

Facilities Breakdown Structure

The Intego programme execution approach as described in Chapter 4 is again presented here as Figure 5.2 to illustrate the scope of the programme. The first round facilities breakdown structure for the Intego programme, based on the scope as shown in Figure 5.2, is presented as Figure 5.3.

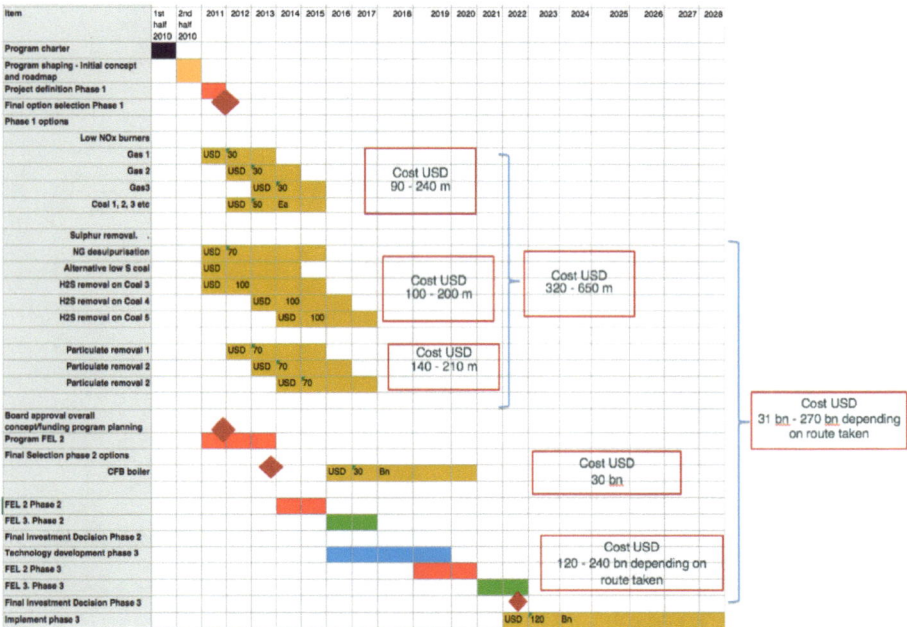

Figure 5.2: Intego programme execution approach

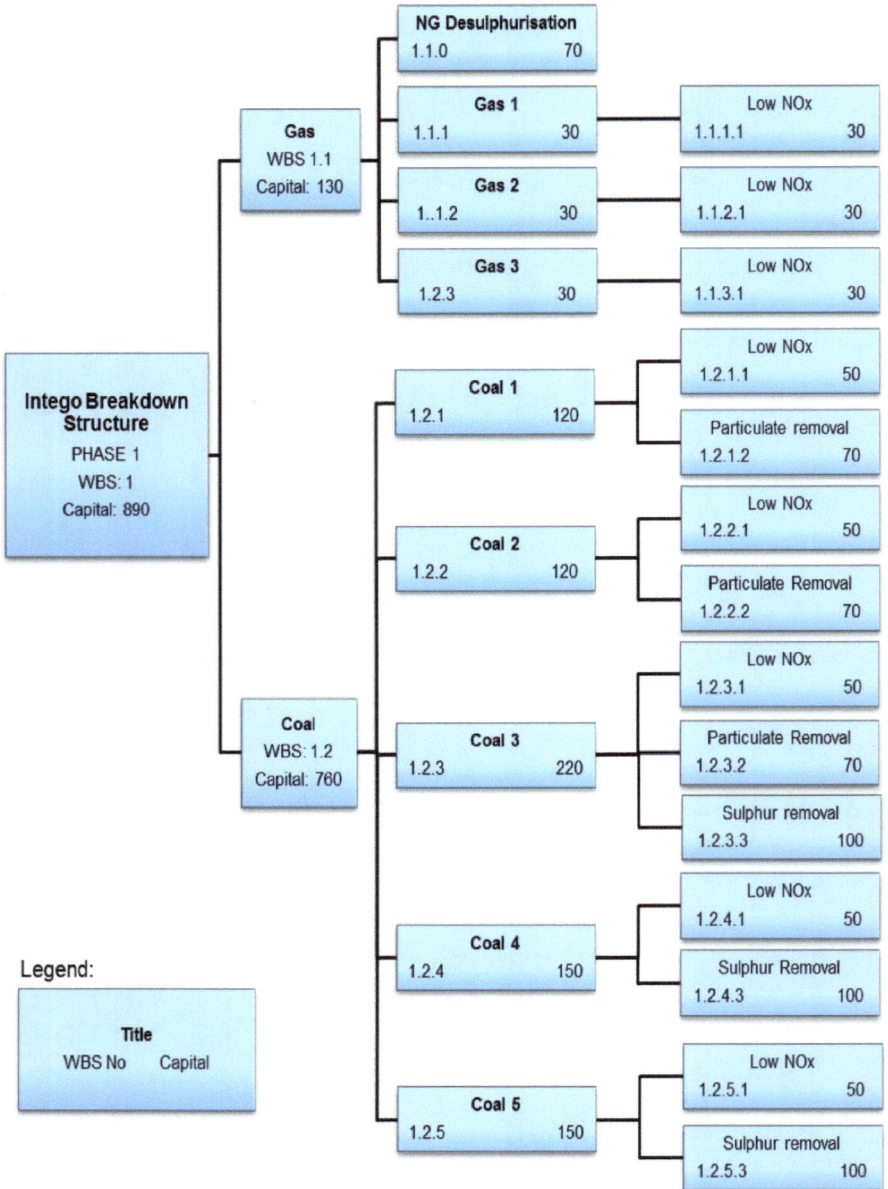

Figure 5.3: Intego facilities breakdown structure – first iteration

Figure 5.3 indicates the overall intent in the form of a facilities scope. The first round schedule could be drawn up based on this work breakdown, but one should have one more look at the breakdown structure and consider if there are any interdependencies between the work packages that could dramatically influence the schedule. The programme execution plan could also result in changes in the schedule. At least one more iteration is recommended before the more detailed schedule is communicated, in order to ensure it will reflect reality as best as possible at this early stage. For the purpose of this discussion it is assumed that there is no interdependence between the different power stations themselves, but only within each power station.

Coal 3 will be used to show how the next revision could look. The interdependencies map in Figure 5.4 shows that certain utilities such as instrument air and an upgrade of the control room will be required for all three emission reduction processes. Cooling water will be required for two of the steps and demineralised water for the low NOx burners. It may therefore make sense to develop these utilities for the complete Coal 3 power station, rather than separate for each unit. This reduces unnecessary duplication in utilities and results in a lower capital expenditure

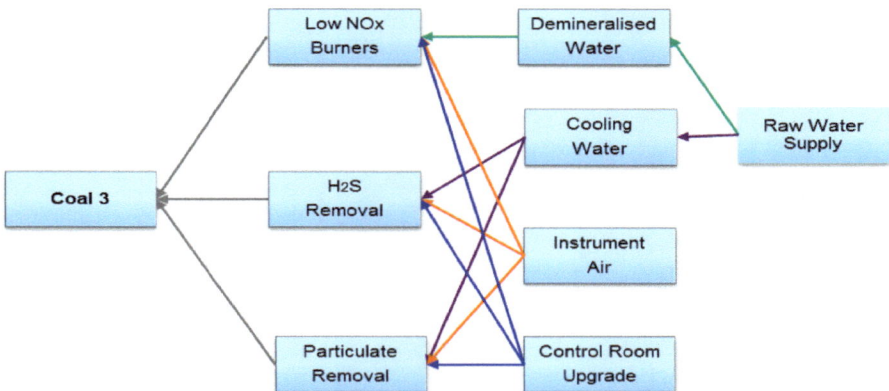

Figure 5.4: Coal Station 3 interdependencies

Considering all interdependencies as illustrated for Coal 3 in Figure 5.4, the second iteration of the facilities breakdown structure is developed, as shown in Figure 5.5.

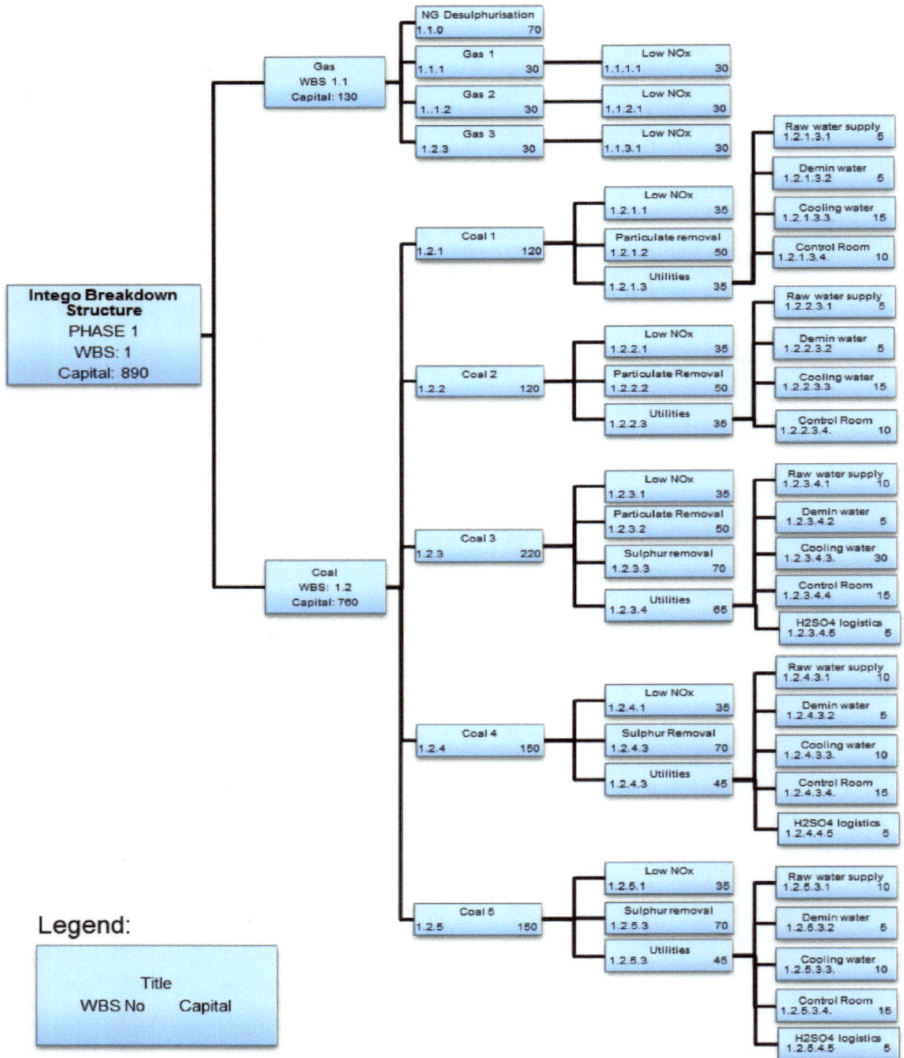

Figure 5.5:   Expanded scope to indicate effect of interdependencies

This second diagram more accurately shows the scope in a way that it can be reported on and controlled, as it is more aligned to the way the individual projects will be executed.

Note that the total cost for each power station has not increased, but the utility allocations (assumed to be 30%) within each package have been extracted into the new utilities package for each of the power stations.

As can be seen from Figure 4.2, representing the master schedule as presented in the programme charter, typical very high level project duration best guesses were used to develop this initial view. In order to verify this timeline, more analysis is required. It is important to agree some assumptions on which to base the plan.

Assumptions

The following are examples of assumptions required to complete the facilities breakdown structure:

- Present stage-gate status of each project (i.e. is it only an idea, has a basic concept been developed?)

- What execution strategy will be followed for each project?

- Will the initial studies be executed by internal personnel or contractors that need to be appointed following a bidding process?

- Will the further development be bid on from a fixed price basis or reimbursable basis?

- What will the bidding and award timelines be?

- How long will each phase be (typical benchmark schedules based on the type of project, the size and the complexity)?

- Which projects must be completed before others, based on interdependency maps?

Once the specific schedule assumptions have been agreed for the projects, a more detailed 'target' schedule can be developed showing the expected duration of each project phase, the interdependencies and total programme schedule. Capturing these assumptions from the start and keeping track of their validity during the course of the programme is a key success factor.

All this work is done at the master schedule level to get an initial understanding of the scope and schedule to aid planning, scheduling and organising of the different resources required to execute the programme. The intent is not to go into level 2 and 3 type of planning at the programme level. The risk of not developing the initial project planning at a reasonable level of detail is that unrealistic schedules, costs and thus stakeholder expectations are created. These 'misconceptions' will only become apparent when the individual project teams deliver their detailed plans, which can be very late in the overall development of the programme.

Capturing and managing project assumptions from the start and keeping track of their validity during the course of the programme, is a key success factor.

Some key assumptions

As a start it is assumed that for the phase 1 projects, the project definition will be completed by in-house resources. The work packages will be submitted for external competitive bidding for the design development and engineering phases on a reimbursable basis. The project execution will be bid on a lump sum fixed price basis up to ready-for-operation, after which an owner team will start-up the plant.

With this information a more defined framework plan or master schedule can now be developed. This will assist in establishing the key milestones and strategies to follow in order to best meet the requirements, as defined in the programme charter.

Programme scheduling

The next round of planning can be done using spreadsheets (e.g. MS Excel, Numbers) as was done in the diagram presented above, but more advanced tools exist that simplifies this planning process. In practice, two types of planning tools are available, namely detail project execution planning tools (e.g. MS Projects, Primavera) and tools that focus on 'management' planning. These latter tools typically do not require complicated logic in order to present a master schedule and work more like a drawing board. Plans can be developed from high level assumptions and metrics to present overviews for management review (e.g. Milestones Professional).

We have found that using what we call 'management planning tools' are much better suited to this initial planning phase. They are also able to present a much better overall view of the programme and the key milestones and can be used in alignment discussions or presentations to steering committees and boards. The two types of planning tools can also be used interactively and the data can be exported and imported between the two, in order to satisfy the needs of both management and the project teams.

Programme scheduling involves much more of a top-down and bottoms-up iterative approach than the planning of a single project, even at the master schedule level (level 1 schedule). The suggested methodology is to start with the overall high level schedule as proposed in the programme execution approach to ensure that the management vision and high level concepts are fed down to the programme, i.e. a top-down approach. Furthermore, a bottoms-up approach using detail planning tools to develop the individual project schedules and integrate them into a feasible programme master schedule. The resultant master schedule (or roadmap) then needs to be compared with the overall 'desired' business schedule and the gaps identified. The assumptions made in developing both schedules can then be reviewed and aligned until the overall master schedule (or framework plan) and the business needs, match to the best extent possible.

An essential element during the planning process is the need to identify and map all interfaces between the different projects within the programme, as well as with the existing operations that may affect the successful outcome of the programme. This assists in highlighting the critical path from a programme perspective.

It is also important to be able to communicate the master schedule (or master plan, depending on your preference) effectively. For this purpose it must be realised that the executive programme sponsor and steering committee require an overall 'executive' view of the schedule. The programme director will have to report progress against this master schedule (a level 1 schedule), whilst the individual project teams operate with levels 3 and 4 type schedules. The communication needs of all parties must be met effectively and the level of the schedule must be appropriate for the communication requirements.

We now return to the ongoing Intego case study to illustrate the programme scheduling process.

Intego Environmental Compliance Programme (IECP)

Programme scheduling

We will start by developing the overall schedule for the conversion of the gas fired power stations to low NOx burners. The assumption is that these conversions are fairly simple and only require the installation of back-mix burners to lower the combustion temperature. No changes to any utilities or control systems are required. The high level summaries of the 3 conversions are shown in Figure 5.6.

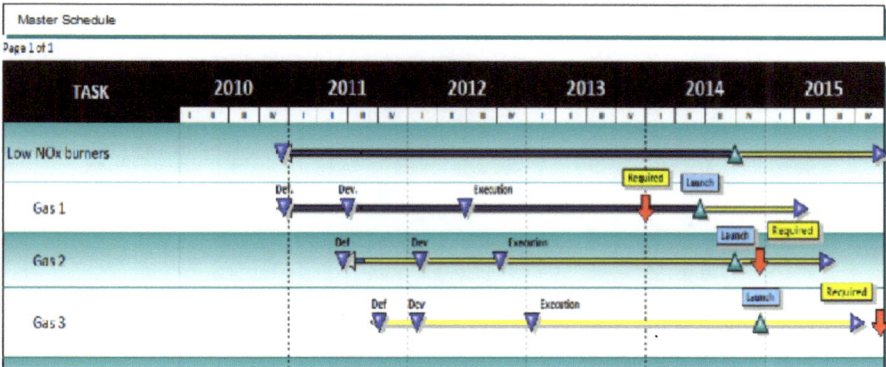

Figure 5.6: Summary master schedules for gas station low NOx burners

The schedules for Gas 1, Gas 2 and Gas 3, as presented in Figure 5.6, are rolled up summaries based on more detailed schedules built up based on the company's stage-gate project development model, taking into account the agreed contracting strategy as well as the required approvals, bidding cycles, etc. As an example, the more detailed schedule for Gas 1 is shown in Figure 5.7. It is important to note that master schedules should be done from an integrated programme perspective and also from a project perspective.

In Figure 5.6, it can be seen that the required dates by business (red down arrow) are generally earlier than what the first project planning indicates (blue right pointing arrow). Especially Gas 1 is about 18 months later than what the business prefers, but it is possible to complete all the conversions 3 months ahead of schedule. The business development and project teams must determine whether this plan is acceptable or agree strategies to close the gap. For the purposes of this discussion we will assume that the framework as presented above has been accepted.

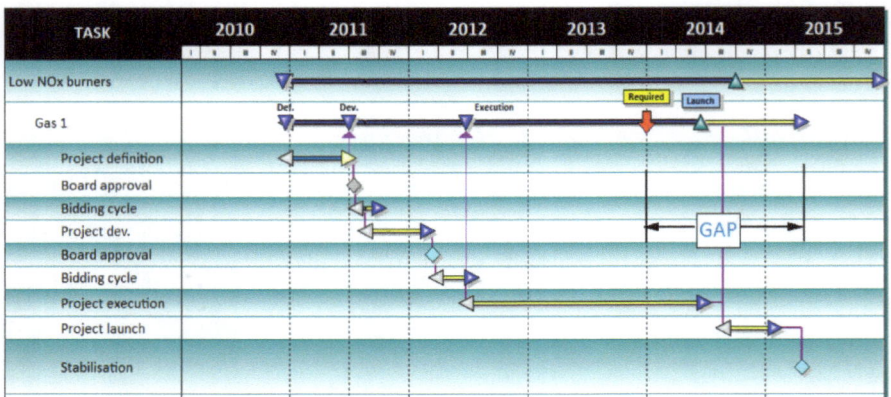

Figure 5.7: Detail master schedule for gas station 1 (Gas 1) conversion

The previous example is relatively simple in that each project was assumed to be independent of the others. The different emission control conversions on the coal power stations are very interrelated and highly dependent on provision of utilities and infrastructure as already indicated in the interdependency diagram in Figure 5.4. Understanding these relationships is crucial in order get an appreciation of the integrated nature thereof and the potential implications thereof on the programme schedule.

It can be seen from Figure 5.4 that various other systems need to be installed and be ready before the actual emission control unit can be delivered for the coal-fired power stations. For example, the control room must be upgraded before any other system can be commissioned, a raw water supply contract must be negotiated and a supply line must be installed and cooling water must be provided. It was decided that as far as possible, the infrastructure upgrades required will be developed to supply the needs of all projects for each power station. In addition, H_2S removal cannot be commissioned before low NOx burners are fully operational. Both low NOx burners and H_2S removal equipment must be operational before the removal of particulates can be done.

Because of the limitation on available construction manpower, it is not possible to overlap construction activities on Coal 1, Coal 2 and Coal 3 for similar projects by more than 50%. Figure 5.8 is thus the best-estimated timeline for conversion of Coal 3, including all the prerequisites regarding utilities and phasing of projects. Taking the above into account, the final completion of all conversions on Coal 3 is between 9 months to 33 months later than originally anticipated.

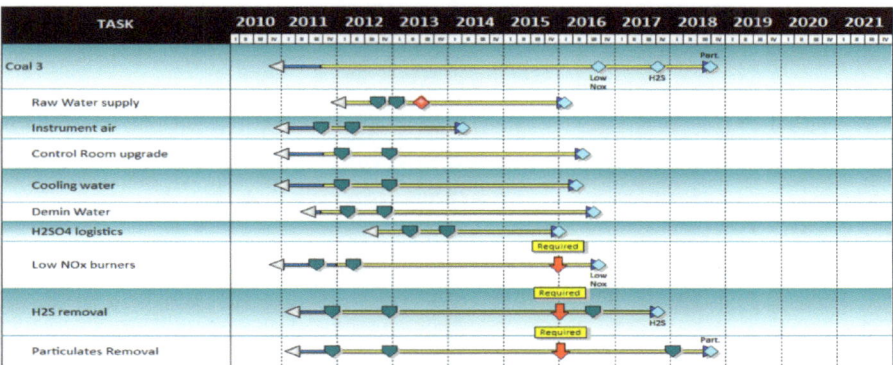

Figure 5.8: Master schedule for the Coal 3 conversions

The total conversion of all three coal-fired power stations, as indicated in Figure 5.9, is predicted to be completed by the end of 2020 as compared to the original requirement of end 2017 for the total phase 1. Alternative strategies for accelerating the schedule, by, for example, training more construction resources, different contracting strategies or engaging with the authorities for a time extension, may be required in order to reach an acceptable overall schedule.

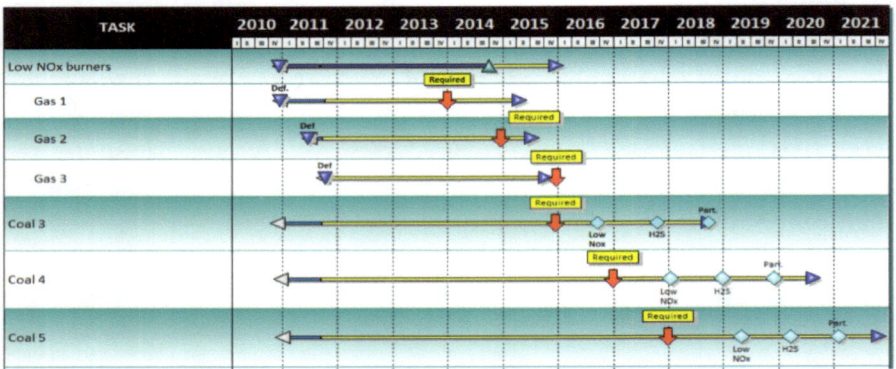

Figure 5.9: Master schedule for coal station conversions

Agreement and project planning

Agreeing the overall timeline and execution strategy will require extensive discussion and alignment with the stakeholders of the programme. Strong leadership from the programme management team as well as facilitation, negotiation and alignment skills will be essential elements to secure an acceptable solution for all parties.

Once the overall strategy and timeline has been agreed, the high level schedule as presented in the Intego case study can now be given to the individual project teams for compiling a much more detailed plan using traditional project planning tools.   All discrepancies with the agreed master schedule need to be highlighted to the programme management team and action taken to address the issue. Alignment on the integrated programme master schedule is critical and requires an iterative process. The assumptions underlying all schedules and the master schedule should be continuously defined, monitored and updated when required.

The necessity for this amount of planning at this early stage cannot be stressed enough as the final cost estimate and schedule that the project team commits to, relies heavily on this level of detail. It will prevent unrealistic promises to boards or enable the project team to develop alternative strategies to satisfy the overall project drivers, be it cost or schedule.

Once the project plans are agreed and approved, it is up to the individual project teams to execute the work according the plan. The programme planners must roll-up the project progress reports into the overall programme progress reports. This should be done against the master schedule for the programme.

As an absolute minimum, the programme planner should ensure that a coding structure is developed and implemented across the whole programme which addresses the:

- **Work Breakdown Structure (WBS):** The WBS, or programme scope, has been discussed extensively in this chapter;

- **Organisation Breakdown Structure (OBS):** The OBS is a coding structure which facilitates the correct usage and supply of resources, primarily people, and their performance, and;

- **Code of Accounts (COA):** The COA is typically developed and maintained by the estimating department. A COA contains a list of prime accounts within which the project work is done. These prime accounts are further sub-divided into sub-prime accounts and sub-prime accounts are further detailed into detail accounts. Each sub-division is a further detail of the previous account. When used correctly, it provides very useful information for estimators regarding costs, durations, etc.

By integrating all of these codes at the work package level, the planner is in the extremely fortunate position of being able to extract information and generate a multitude of reports of various types. It does not mean that these are the only codes required, but, these are the codes that are required for proper planning and control of any project.

Typically a Cost Breakdown Structure (CBS) is also required; this allows for the project budget to be integrated and controlled via the schedule. However, it becomes meaningless if costs are not integrated into the schedule.

Alignment on the integrated programme master schedule is critical and requires an iterative process. The assumptions underlying all schedules and the master schedule should be continuously defined, monitored and updated when required. The integrated master schedule should clearly define project sequence and interdependence, and must be reported and managed accordingly.

Work package scope definition

The programme planning process requires one more step in order to document it fully and ensure alignment amongst all stakeholders. Each of the summary work packages, as defined in Figure 5.5, needs to be documented and approved.

A work package needs to be described at a sufficiently high level of detail such that it is clear to the eventual business owner what he or she will receive and to the engineering team what they should deliver. It is important that each work package description is accepted and signed off by these two parties (business and engineering). Acceptance of the work packages is a crucial part of the shaping process.

A work package description is not a detailed engineering document and therefore should not go into too much engineering detail. It should be at a level such that the business owner can clearly understand the scope to be delivered, bearing in mind that he or she may not be a qualified engineer or have the detailed engineering understanding of a project professional. It should, however, be at a sufficient level of detail that project engineering can also use it as the starting point for further definition of the work package.

Typical engineering deliverables such as process and mechanical flow diagrams, instrumentation diagrams, engineering data sheets, etc. should not be part of the summary work package description. Simplified block flow diagrams or similar high-level representations can effectively be used to ensure that non-technical people fully understand the concepts.

A scope definition per work package typically includes the following items:

- Header box containing the project title, WBS name, WBS number;

- Key milestones dates and anticipated cost;

- One line description of the objective of the work package;

- Summary of key items included in the work package;

- Summary of the key items excluded in the work package, and;

- A battery limit diagram indicating the battery limits for the specific package.

We shall now return to the Intego case study to show how work packages are defined.

Intego Environmental Compliance Programme (IECP)

Work package scope definition

Project:	**Coal 3 - Environmental Compliance Project**
Work package	**Utilities**
WBS Code	1.2.3.4
Key Milestones	Refer to master schedule included as Figure 5.10
Capital/Operating cost	• Capital cost $65m • Operating cost assumed as 3% of capital
Revision number	1.0
Date	1 March 2010
Approved by Business Owner	Signature
Approved by Project Manager	Signature

Objective

To provide all utilities required for the modifications to Coal 3 power station as envisaged in phase 1 of the IECP.

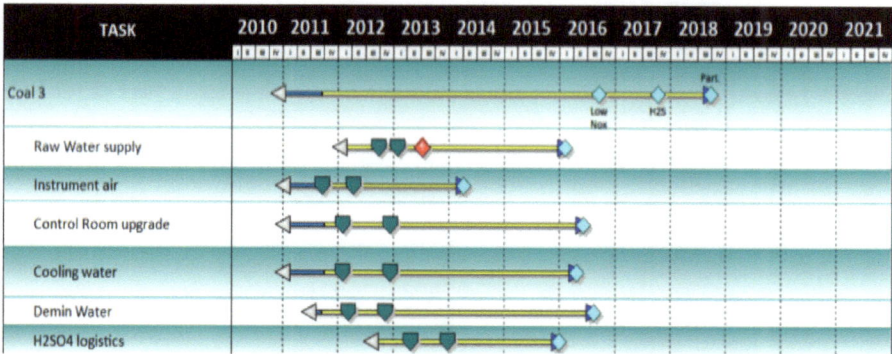

Figure 5.10: Master schedule for Coal 3 power station

Included in scope

The scope of the Utilities package for Coal 3 includes:

- Raw water intake from river and purification for cooling water and plant utility water purposes, including intermediate and final storage;
- Central cooling tower including side stream filtration, chlorination and corrosion management system;
- Plant and instrument air system including compressors, accumulators and driers;
- Electrical power supply including connections from the main plant substation as well as the required extensions to the main substation and the reticulation to the substations for the other work packages, and;
- The scope includes all the reticulation cables and pipe work from the common utility area up to the battery limit of each of the consuming packages.

Excluded from the scope

The scope of the Utilities package for Coal 3 excludes:

- Utility supply requirements for future phases (phases 2 and 3) of the IECP. The capacities of the different utility units will only cater for the needs as envisaged in phase 1 of the programme. No pre-investment will be made for future expansion, either in plot area required or in the sizing of any of the hardware components, and;

- Nitrogen supply as a common utility. Should nitrogen be required by any of the systems, provision must be made within the scope of the particular package.

Battery limit diagram

Battery limit diagrams are required for each of the utilities listed as included in the scope of the work package. For the sake of brevity, only the diagram for the water systems is included as Figure 5.11.

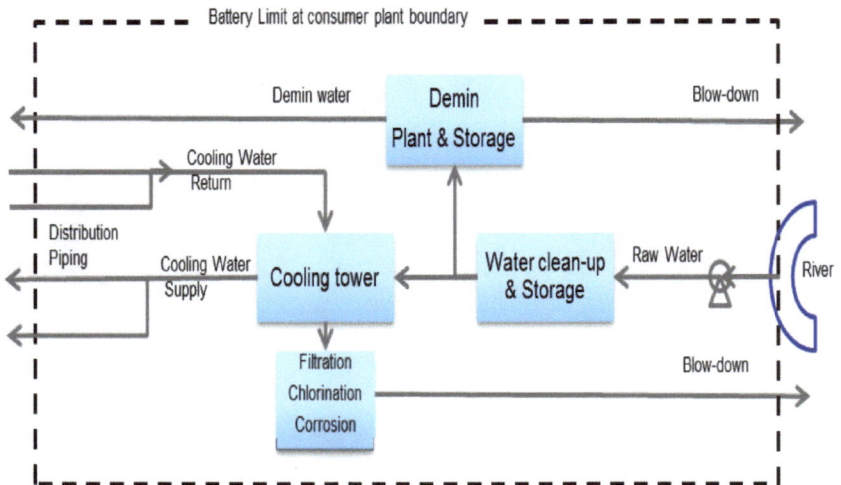

Figure 5.11: Battery limit diagram for water systems

Interface management

Whenever there is more than one work package, the need for proper interface management exists. As the definitions of scope of the various packages are being completed, it is the function of the interface and integration management team in the programme office to ensure that each package is defined clearly. This implies that no gaps or overlaps exist between the packages. Gaps mean that some of the scope could fall through the cracks, only to be discovered later and thereby causing very expensive rectification. Overlaps imply duplication of equipment and services, with similar cost and rework implications

It is recommended at this stage to review the capital cost estimate to ensure it actually includes all the scope as defined in the scope definition package before final sign-off and alignment.

It must be pointed out that all the above planning is based on level 1 type 'generic' information. Typical scopes for packages, benchmark costs and durations are used. Note that no actual engineering work is done as yet.

Interface management means that no gaps or overlaps exist between individual work packages. Either of these is undesirable because it carries cost and rework implications.

Concluding remarks

The objective of this initial planning phase is to clearly understand the implications of embarking on the programme with its associated projects and ensuring full alignment of all stakeholders before embarking on developing the engineering work. Diligently completing the high level planning ensures that once you embark on the actual programme (that is now fully understood and properly controlled), the probability is very high of completing it within the agreed cost and timeline and meeting the overall business objectives.

As a final closing remark on the planning section, it must be stressed that this book focuses on the role of the owner project management team. The owner team must develop the full specification for the project that will meet the business needs and shareholder expectations. Once the specification is complete, engineers and technicians are appointed to do the engineering design, procurement and construction of the facility. The owner team needs to manage these engineers, whether they are from an outside contractor or from in-house resources, to ensure that the project will be delivered as expected. It is not the role of the project owner team to actually do the engineering work.

In the next chapter we focus on programme organisation and the owner project management team.

Chapter 6:
Giving Shape to the Programme Organisation

"First comes thought; then organisation of that thought into ideas and plans; then transformation of those plans into reality. The beginning, as you will observe, is in your imagination." - Napoleon Hill

Introduction

Once the overall context of the programme is envisioned and approved, the next step is to set up the overall owner programme management organisation, as indicated in Figure 6.1. This implies defining the respective roles and responsibilities of the programme team and also what the respective project teams will be responsible for.

In addition, a programme office needs to be created with the necessary infrastructure and project controls systems to enable sound control and governance of the programme. The required communication tools and protocols need to be developed and implemented to ensure that all stakeholders are kept in the loop and that the necessary reviews and approvals are obtained in good time.

A note of caution at this point is required. The structures discussed below ideally reflect the owner organisation and not that of the engineering contractor, the project management contractor, the managing contractor or any other such organisations. In many instances owners do not have the required competencies or resources to fulfil all the roles as laid out in this chapter. As an absolute minimum, owner personnel should occupy the lead roles and be accountable for the core owner team activities. This responsibility cannot (and should not) be delegated to an engineering contractor. This holds true even if the engineering contractor and owner personnel work as an integrated programme management team. These activities are elaborated on further in this chapter.

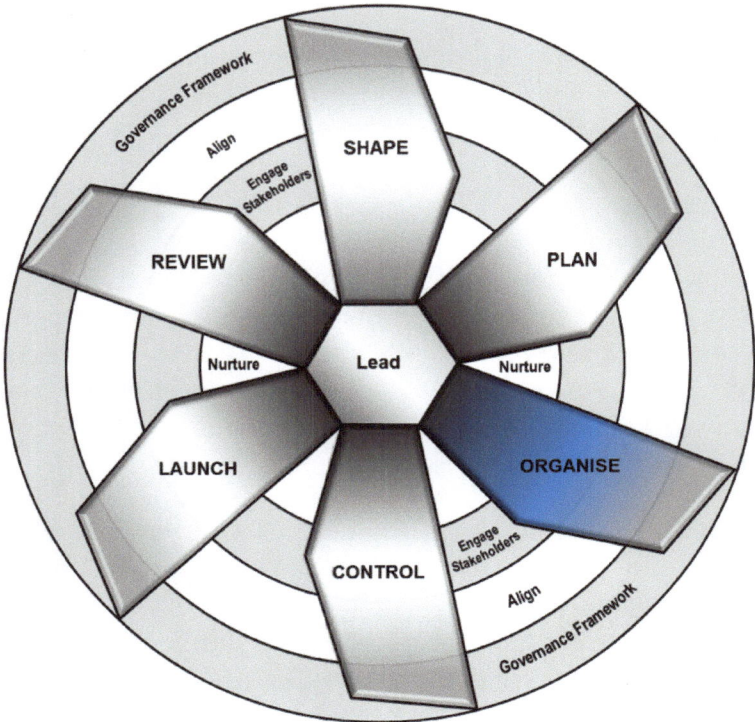

Figure 6.1: Programme management model with 'Organising' highlighted

Owner organisation personnel should occupy the lead roles and be accountable for the core owner team activities in a programme. This responsibility cannot be delegated to an engineering contractor. Contractors cannot be a substitute for owner teams.

The question remains as to why an owner programme management team is required and why so many resources are necessary. Merrow (2011) answers these questions in his recently published book *Industrial Megaprojects*. According to Merrow, the role of the owner team is to generate comparative advantage for the owner organisation. The owner team is where all the owner functional groups come together to take a business opportunity and define a project that is fashioned to the particular strengths and talents of the owner organisation.

The development of the scope and preparation for execution of a programme requires the involvement and active participation both inside and outside of the owner organisation. Merrow continues that it would be easy to assume that gaining cooperation from outside organisations is problematic, but often it is more of a problem to get support from all functions inside the organisation. Different functional groups in organisations have different reporting lines, different measures of success, and different concerns. The owner programme team has to integrate the services of many functions to create a programme that meets the business objectives of the organisation.

This integration role cannot be performed by outside contractors for a number of reasons, namely:

- Contractors have almost no knowledge of operations;

- Contractors do not understand the business of the companies they serve, and;

- Contractors work for their own shareholders and not those of the owner organisation.

However, Merrow (2011) states that contractors still remain indispensable for the execution of programmes and megaprojects. The owners that do essentially all of their project work in-house have to rely on contractors to execute their megaprojects, but contractors cannot be a substitute for owner teams.

Setting up the necessary project controls and communications are addressed in more detail in following chapters. This chapter will focus on setting up the programme management team and defining the roles and responsibilities of the members and also setting up the programme office.

Establishing a programme management team

Introduction

In Chapter 1, we defined programme management as the co-ordinated organisation, direction and implementation of a group of related projects and activities that together achieve outcomes and realise benefits that are of strategic importance. Managing multiple related projects by means of a programme allows for optimised or integrated cost, schedules, integrated or dependent deliverables across the programme, delivery of incremental benefits, and the optimisation of staffing in the context of the programme's needs.

Specific synergies are achieved by the definition of a number of projects which collectively deliver the overall business objectives. These projects are defined and instigated by the programme team, but will have their own project management teams with the programme manager providing overall guidance only.

Once a project is started, the programme manager should not intervene directly, but will need a degree of feedback and control. The project manager is concerned with project-level information e.g. progress, issues, risks, costs, projected benefits, etc. It is unwise to feed all such data to the programme manager. In principle, only those items affecting the overall programme need to be communicated.

Programmes and projects often deliver substandard results because teams are not set up and structured properly. Setting up teams implies that all the necessary functionalities required for success are represented on the team and that each member understands the business and programme or project objectives (programme or project charter) as well as his specific role and responsibility. All members should understand the key programme or project objectives, master schedules and deliverables and ensure that risks are identified and mitigation steps are being followed up.

Competencies required

A programme management team differs from a normal project team only in that the team members need to understand and focus on the overall programme objectives. The programme management team must direct the individual project teams towards those goals rather than control the individual projects on a day-to-day basis. All the functionalities normally required on a project team must also be represented on a programme management team. In addition, it is possible to set the organisation structure such that common resources across all projects form part of the central programme team and provide their services to all the projects within the programme on an as-required basis. As an example, human resource management require more focus and planning because of the long duration of a programme where individuals on the team still need personal development, promotions, and administrative support. Continuity of the team resources is critical on any project and succession planning therefore becomes a real issue on programmes.

In order to better understand the functions of a programme management team, examples of key competencies that are required within an owner's project management team are listed below (CII, 2014):

- Setting project goals, objectives and priorities;

- Project management oversight;
- Performance metrics and benchmarking;
- Conceptual planning and road mapping;
- Project controls, planning, progressing and scheduling;
- Procurement and supply agreements;
- Constructability and construction management;
- Safety management;
- Risk management;
- Legal and contract administration;
- Preliminary design and scope development;
- Conceptual design;
- Detail engineering;
- Total quality management;
- Conceptual and definitive cost estimating;
- Health risk management;
- Government and public relations;
- Business development;
- Finance and budgeting;
- Maintenance and operability;
- Commissioning and start-up;
- Performance testing;
- Human resource management;
- Team development and wellness;
- Communication;

- Contracting, alliancing and partnering, and;

- Environmental management and permitting.

When a project or programme management team is mobilised, it is recommended that a list of competencies required for the programme/project is developed. These competencies should be classified according to whether it is a core or a supportive competency, a nice-to-have or perhaps not required after all. In setting up the team, it must be ensured that all the required competencies (core and supportive) will be represented in the owner team. It cannot be assumed that a team will automatically possess all the required competencies. It must also be realised that every individual on a project team has a unique set of competencies and the team set-up, roles and responsibilities need to be specifically developed for each team. Responsibilities of members of the team should be structured with the individuals' strengths and weaknesses in mind.

We return to the Intego case study to see a worked example of competency analysis and the programme management team organisation structure.

Intego Environmental Compliance Programme (IECP)

Programme Management Team Organisation Structure – Part 1:

The competency analysis in Table 6.1 was specifically developed for the Intego programme to illustrate the key competencies required.

Table 6.1: Competency analysis Intego programme

No	Competency	Category	Comment
1	Setting project goals, objectives and priorities	Core	Key functionality of the programme management team
2	Project management oversight	Core	Key functionality of the programme management team
3	Benchmarking & metrics	Supportive	Enables team to set realistic targets and compare results
4	Conceptual planning/ road mapping	Core	Key functionality of the programme management team
5	Project controls, cost, planning, progressing & scheduling	Core	Required to understand and roll up detailed plans, progress, costs and forecasts to programme level
6	Procurement	Core	A centralised procurement group at programme level in order to distribute orders and manage vendors in a consistent manner will handle all programme procurement
7	Constructability & construction management	Supportive	Constructability and construction management oversight will be at programme level but each project will manage its own construction
8	Safety management	Core	Safety will be directed from programme level with consistent approaches and directives
9	Risk management	Core	Standard risk management procedures of the company will be used
10	Financial approvals	Core	As per company guidelines
11	Legal/contract administration	Core	All programme contracts will be managed at programme level
12	Preliminary design/	Core	Required in order to

No	Competency	Category	Comment
	scope development		conceptualise the overall programme
13	Conceptual design	Core	The programme team needs understanding in order to ensure overall integration, but concept design will be done by each project team
14	Detail engineering	Not required	Will be done by subcontractors (see quality).
15	Total quality management	Core	Programme will set quality standards and guidelines, but each project team needs to implement specifically for the project
16	Conceptual & definitive cost estimating	Core	Conceptual cost estimating will be done by projects and programme team but definitive estimates by subcontractors with owner overview
17	Health risk management	Supportive	Guidelines for health issues only, but handled by each project
18	Government & public relations	Core	Key to the success of the programme is reaching an overall acceptable plan with all stakeholders
19	Business development	Core	Key to shaping the correct programme
20	Maintenance/operability	Core	Guidelines at programme level. Each project to manage
21	Commissioning/start-up/performance testing	Core	Guidelines at programme level. Each project to manage
22	Human resource management	Core	Essential to ensure right competencies, continuity and retention of personnel

No	Competency	Category	Comment
23	Team development	Supportive	Requires attention in order to ensure teams understand the overall programme and project objectives, their respective roles and responsibilities and work as an aligned team
24	Communication	Supportive	On such a large and long-term programme communication is especially important
25	Contracting/Alliancing/ Partnering	Supportive	Required in order to agree the overall contracting strategies and management thereof
26	Environmental management & permitting	Core	In order to set correct objectives and timelines and achieve required environmental permits in time
27	Lessons learnt transfer and best practices	Supportive	Required in order to enhance overall programme effectiveness

From Table 6.1 above and considering the respective competencies of the prospective team members, the following programme management team has been developed for the Intego project. The number of the respective competency (as reflected in Table 6.1) is indicated for each position/person in the organogram. The first round organogram may, for example, only focus on the core competencies (indicated in red in Figure 6.2) and resourcing of key personnel with the required core competencies. Once these personnel have been appointed, it can be ascertained what additional competencies may be present in this team. Additional positions may then have to be created to completely satisfy the competencies required for the IECP.

The final programme management organisation structure may be as illustrated in Figure 6.2, with the number referring to the competencies as per Table 6.1.

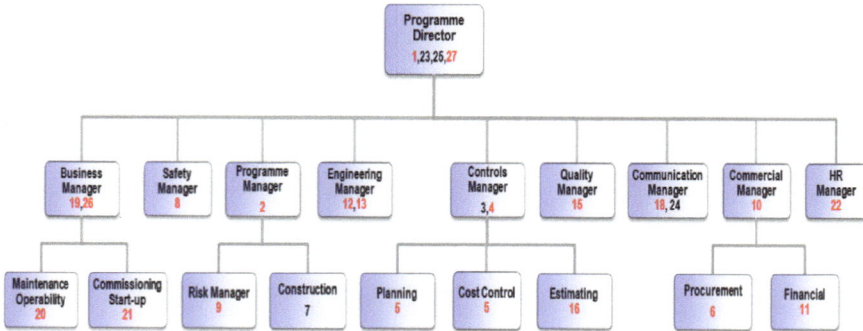

Figure 6.2: Programme Organisation Structure

Finalising the organisation structure

The following discussion should be read with the mind-set of understanding the thought process, rather than the exact structures, roles and responsibilities presented.

The programme management team is accountable for the overall successful execution of the total programme with the single-point responsibility being the programme director. The programme director, in turn, reports to the executive sponsor for the programme who is responsible for the business outcomes. As can be gathered from the discussion, finalising the organisation structure is an iterative process. Once specific key individuals in the programme leadership team have been appointed, it may be required to adjust the initial organisation structure to match with the competency profiles of these individuals.

As already mentioned in Chapter 1, one of the key reasons to set up a programme management structure rather than execute each project separately, is to manage a group of related projects in a coordinated way to obtain benefits and control not available from managing them individually. Some of these benefits are the ability to allocate dedicated common resources to the programme that will not normally be possible on an individual project basis. The skills acquired and lessons learned can then be easily applied from project to project and resources can be applied much more effectively. It is therefore required to consider which competencies will be made available to the project teams from a common programme pool and which must be dedicated to each project. A specific complicated sub-project may also require additional dedicated resources rather than using common team resources. At the outset, it has to be established how this programme concept will be implemented in practice and what resources and competencies will be established as a common pool and what will be dedicated to individual projects.

We return to the Intego case study to finalise the organisation structure for the programme management team.

Intego Environmental Compliance Programme (IECP)

Programme Management Team Organisation Structure – Part 2:

Competency Table 6.1 above has been expanded on a generic basis to indicate which competencies will be dedicated to project teams vs. supplied on a programme basis. The basis used was to consider the typical type and complexity of the sub-projects of the Intego programme. This provides a starting point, but specific project teams could have a different composition, depending on the needs of each project.

Table 6.2: Competency allocation project vs. programme level

No	Competency	Programme/project level competency
1	Setting project goals, objectives and priorities	Programme level
2	Project management oversight	Programme level for oversight. Each project to have dedicated project manager
3	Benchmarking/metrics	Programme level
4	Conceptual planning/road mapping	Programme level
5	Project controls, cost, planning, progressing & scheduling	Project level but rolled up oversight at programme level. Common programme resources
6	Procurement	All procurement handled centrally at programme level. Common programme resources
7	Constructability & construction management	Project level but overall guidelines and coordination at programme level
8	Safety management	Project level but overall guidelines and coordination at programme level
9	Risk management	Project level but overall guidelines and coordination at programme level
10	Financial approvals	Programme level
11	Legal/contract administration	Programme level
12	Preliminary design/scope development	Project with oversight at programme level
13	Conceptual design	Project with oversight at programme level

No	Competency	Programme/project level competency
14	Detail engineering	Outsourced
15	Total quality management	Programme level with specifics incorporated for each project
16	Conceptual & definitive cost estimating	Programme level
17	Health risk management	Programme level
18	Government & public relations	Programme level
19	Business development	Programme level
20	Maintenance/operability	Project level
21	Commissioning/start-up/performance testing	Project level
22	Human resource management	Programme level
23	Team development	Programme level
24	Communication	Programme level
25	Contracting/alliancing/partnering	Programme level
26	Environmental management and permitting	Project level with guidelines and oversight at programme level
27	Lessons learnt transfer and best practices	Programme level with input from projects

For each of the sub-projects, a specific project team will be appointed (at the appropriate time) with dedicated members for project specific core competencies. The programme management team is thus expanded with the individual project team as required. The sub-project teams will be mobilised and demobilised as required by the individual projects.

Figure 6.3 is a more developed organisation chart and shows the individual project teams with the programme supporting structure above. The engineering and commercial disciplines will support all projects from the programme level.

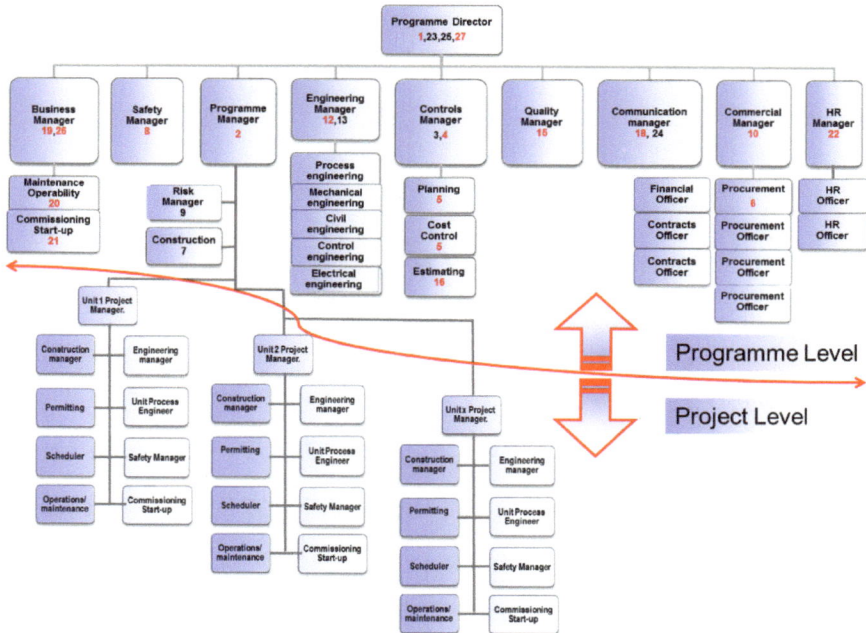

Figure 6.3: Expanded programme and project organisation structure

The programme functional managers will be accountable for their function for the programme and the project functional managers will take guidance from the programme functional managers. For example, the safety manager will be overall accountable for the safety of the programme, whilst the project safety managers will take guidance from the programme safety manager. The individual project manager will however remain accountable for safety on his or her project. This inevitably leads to a matrix reporting structure.

The structure depicted in Figure 6.3 above seems to indicate a very strong hierarchical structure, which can sometimes be misleading. It is also good to indicate the matrix type organisation as depicted in Figure 6.4, as this is the manner in which it will actually be functioning.

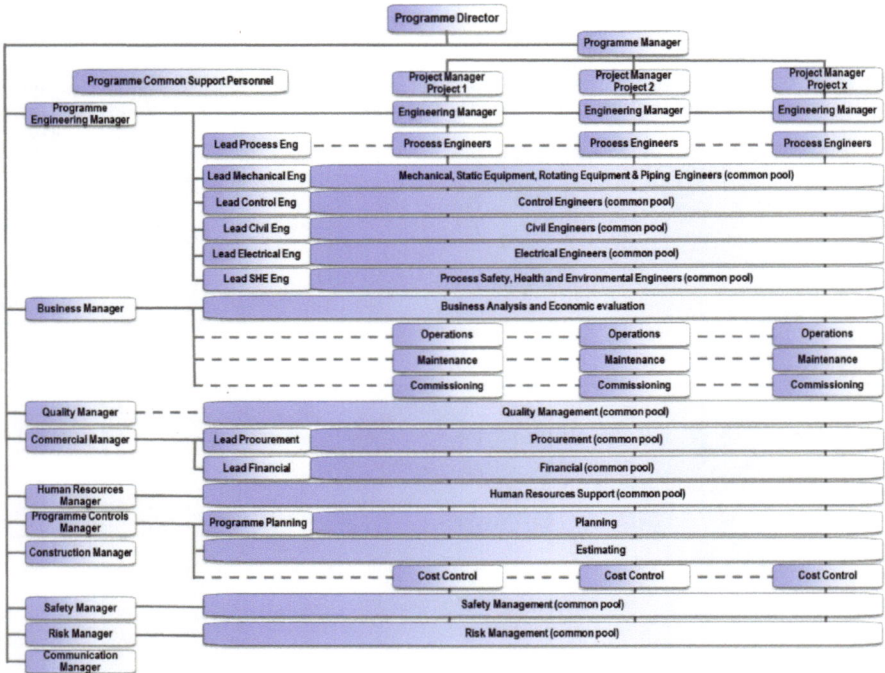

Figure 6.4: Programme and project matrix structure

The programme director is responsible for the overall success of the programme and all sub-projects from an execution perspective. Supporting the programme director is a programme manager and project managers responsible for their specific projects. On the left hand vertical side of Figure 6.4 (columns 1 and 2) are shown the different technical discipline leads accountable for discipline engineering quality. The long horizontal bands indicate the common programme resources that will support all the projects in the programme. Following the vertical line for a specific project, the resources solely dedicated to a single project are indicated as narrow vertical bands. In summary, the responsibility for delivering the project outcomes is illustrated vertically while the responsibility for functional engineering integrity and quality is shown horizontally.

The programme management team organisation structure inevitably ends up as a matrix type organisation with project management reporting lines as well as discipline or functional reporting lines.

Matrix structures: a frame of mind

A matrix arrangement often gives rise to stress between the functional and project lines but it also creates the necessary tension to ensure that both the functional and project components are addressed adequately. The relationships need to be managed correctly in order to ensure that balance is maintained.

Bartlett & Ghoshal (1990) in an article entitled *Matrix Management: Not a Structure, a Frame of Mind*, put some perspective on this matter. They state that the most successful companies are those where top executives focus less on the quest for an ideal organisation structure and more on developing the abilities, behaviour and performance of individual managers. Change succeeds only when those assigned to interdependent tasks understand the overall goals and are dedicated to achieving them.

One of the senior executives interviewed by Bartlett & Goshall (1990) is reported to have said that "the challenge is not so much to build a matrix structure as it is to create a matrix in the minds of our managers." The inbuilt conflict in a matrix structure pulls managers in several directions at once. However, developing a matrix of flexible perspectives and relationships within each manager's mind achieves an entirely different result. It lets individuals make the judgements and negotiate the trade-offs that drive the organisation toward a shared strategic objective.

Job profiling

As the structure is being defined and approved (including the required manpower budget) the requirements for each position can be developed and either advertised or candidates sought internally that matches the respective position profiles. It is our experience that project teams are generally staffed by who-is-available rather than who-is-qualified. Another recipe for disaster is to have all the positions defined but many positions remain vacant while the right person is being searched for and probably never found, while the project proceeds anyway. These practices do not contribute to successful project completion.

Programmes and megaprojects are complex and extend over a long period. It is even more crucial in these cases as compared to single projects to have experienced and competent people in leadership positions right from the start. Programmes and megaprojects generally have impacts on the owner company that can lead to the failure of the company if not successfully delivered. Definition of the profiles and requirements for each position and then sourcing of competent personnel for each and every position is not negotiable.

We return to the Intego case study to have a closer look at job profiling. Every job or position in the programme management team organisation structure, or group of similar jobs, needs to be profiled. This is essential to ensure that the best candidates can be sourced for the positions available. Appointment of personnel to the programme happens stage-wise, starting with the programme director and key programme leaders followed by project team members, as per the individual project requirements.

Intego Environmental Compliance Programme (IECP)

Job Profiling

The complete set of job profiles for the organisation structure as presented in Figure 6.2 will be required. Such a set of job profiles will require much effort and will cover many pages. For illustrative purposes, we have selected the position of programme engineering director, whose job profile is shown in Table 6.3 below.

Table 6.3: Profile for Programme Engineering Director

Job Title	Programme Engineering Director	Job Level	Hay 22
Direct Reports	Eng. discipline leads, Unit engineering managers, Discipline engineers	**Reporting into**	Programme Director
Qualifications	**Work Experience**	**Computer Literacy**	**Purpose of the Job**
B.Eng/BSc Hons Degree (Master's degree in engineering management will be an advantage)	Professional Engineering Manager with 10 years' experience in management of large integrated projects	Computer Literate: Fully proficient in Word, Excel and PowerPoint, Microsoft Projects and SAP	Development and implementation of an overall technical strategy to deliver a successful programme

| **Functional Job Profile** |
| **Key Performance Areas and Main Activities** |
| **1. Accountabilities and Responsibilities** |

Accountable for:
- Overall technical integrity of the project;
- Ensuring that capital estimates accurately reflect technical intent, and;
- Ensuring that uncertainties and risks are included in contingencies and risk allowances.

Responsible for:
- The development and implementation of engineering execution guidelines;
- The development and implementation of engineering quality management for all project phases from feasibility, through execution, commissioning and start-up;
- Providing input into developing and managing of realistic schedules for projects;
- Technical risk identification and management;
- The development and implementation of a safety, health and environmental strategy, including environmental impact assessment;
- The collation and management of the overall scope of the project;
- Quality front-end-loading with reference to all technical/engineering aspects;
- Providing a strategy for and ensuring that all technical interfaces are identified and managed, and;
- Managing sub-project engineering managers and providing leadership, support and an aligned and consistent approach.

| **2. Leadership** |

The candidate should:
- Critically review all activities and results on projects for technical excellence and robustness;
- Foster a culture of collaboration and personal growth of direct reports, and;
- Provide coaching and guidance to direct reports.

| **3. Customer and Relationship Results** |

The candidate should:
- Ensure professional working relationships with all relevant stakeholders in the project execution value chain;
- Negotiate deliverables and realistic schedules for all projects;
- Ensure delivery on all projects as agreed with business partners, and;
- Ensure open, honest relationship with business partners to build and maintain trust.

| **4. Management** |

The candidate should:
- Control and manage systems, procedures and performance requirements, and;

- Identify any deviation from requirements to enable corrective actions to be taken.

5. Competencies and Attributes

Functional Competencies	Personal Attributes
Ability to manage multidisciplinary technical/ engineering work from front-end-loading to completion	Proven leadership skills & ability to work in a team
Competent in different contracting strategies.	Robust emotional intelligence
Competent in estimating and cost control	Reliable and trustworthy
Competent in SH&E requirements on projects	Pro-active and self-starter
Capable of compiling the programme scope from various discipline inputs	Ability to work within a diverse culture
Competent in problem solving, negotiation and facilitation	Ability to work under pressure
Capable of identifying, planning and negotiating for multi-functional resources	Ability to work in a changing and ambiguous environment
Capable of aligning, integrating and managing the various stakeholders and discipline efforts on a programme/project into a coherent integrated engineering execution plan	Approachable
Technical knowledge to enable management of the interfaces between the different engineering disciplines, engineering contractors, project vendors and site specific requirements	Knowledge of site construction activities with a focus on quality and integrity
Capable of evaluating and maintaining capital estimates (cost estimation basis, judgement of cost estimates)	
Leadership Competencies	**General Skills**
Ability to establish and maintain professional working relationship with all stakeholders on a project	Good communication and presentation skills (Written and verbal)
Accept responsibility and accountability for project success	Good interpersonal skills
Ability to establish and lead diverse project teams	Sound management judgement and awareness
Ability to facilitate and manage performance	
Ability to motivate and lead direct reports and project teams	

Sourcing programme team members

The job descriptions, as illustrated for the position of programme engineering director in Table 6.3, are used for sourcing suitable candidates. They can also be a starting point for developing specific project or programme roles and responsibilities. Once the team is assembled and the specific competencies and strengths of the members, as well the specific programme requirements are known, the specific roles and responsibilities of the members must be developed and aligned between all parties. This can be achieved in a work session between all the programme team members where the organisational structures and profiles are discussed. Job descriptions suited to the programme needs, and adapted to the attributes of each team member, can then be developed. Do not assume that everybody knows what he or she is supposed to do, but take the effort to achieve alignment on this matter. Alignment practices are discussed in Chapter 13.

As mentioned at the beginning of this chapter, the owner programme team concept has been developed with the assumption that all the resources are sourced from within the owner organisation. However, it is seldom possible to achieve this, since the manpower requirement for large programmes can easily amount to a few hundred people. This will stretch the company resources to beyond what is practically achievable. One of two approaches can be used to overcome this problem, namely:

- **Small owner team and supporting contractor personnel:** This approach involves changing the structure into two layers, the first consisting of a small owner team fulfilling key owner roles plus a second layer of contracted-in labour. The owner team then supervises a managing contractor (MC) or project management contractor (PMC) that supplies the bulk of the personnel and assumes a large portion of the accountability to manage the overall programme. In the case of the appointment of a MC or PMC, the owner team would conceptually be much smaller with the contractor assuming a greater accountability. Even with this approach, it must be ensured that the required core competencies are still retained in the owner team in order to be able to exercise proper management oversight and giving correct guidance, or;

- **Salt and pepper programme team:** The second approach is to form an integrated management team staffed by both owner and contractor personnel. Depending on the capabilities and availability of personnel, the team is staffed on a salt and pepper (mixed) basis from the two organisations. The implication is that positions in the owner programme team are filled with in-house personnel as far as is practicable and the remainder of the positions are filled with contractor personnel. This allows the owner organisation a wider range of suitably qualified personnel for the different positions.

We return to the Intego Environmental Compliance Programme to illustrate the sourcing of suitably qualified team members and the possible organisation structures that can be used.

Intego Environmental Compliance Programme (IECP)

Sourcing Programme Team Members

The approach of using a small owner team and supporting contractor personnel from the MC or PMC is shown in Figure 6.5.

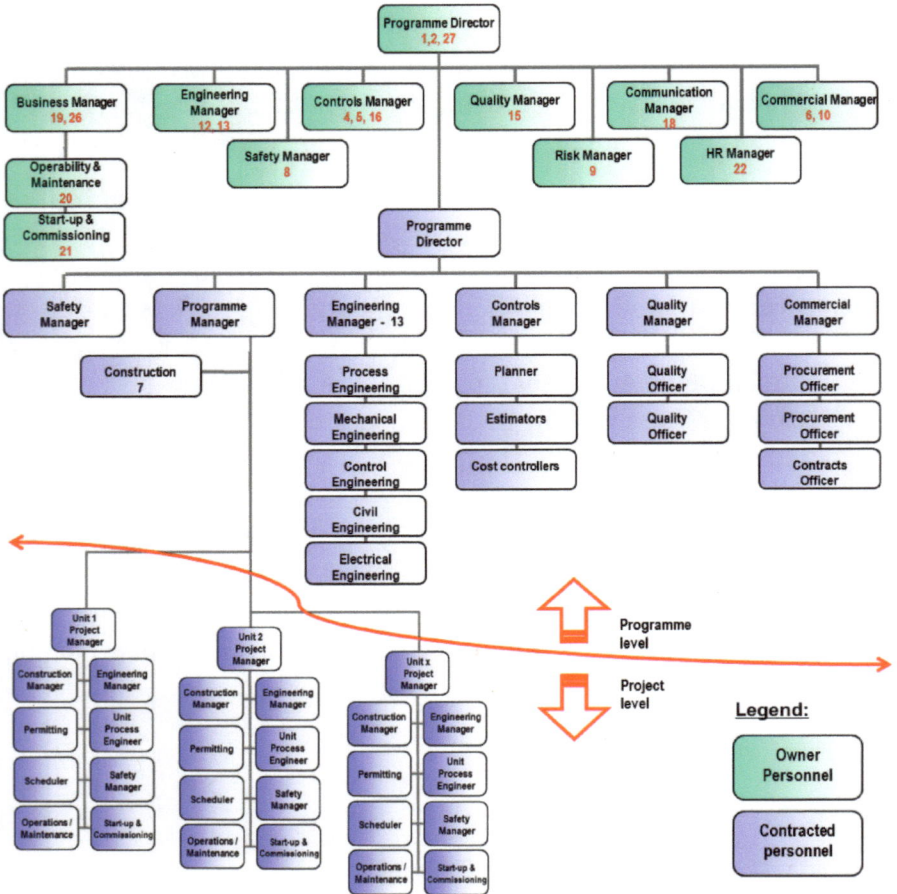

Figure 6.5: Owner team with supporting contractor personnel

The alternative approach to manning the owner's programme team is to use the so-called salt and pepper approach where the team is staffed by both owner and contractor personnel. This approach is illustrated in Figure 6.6.

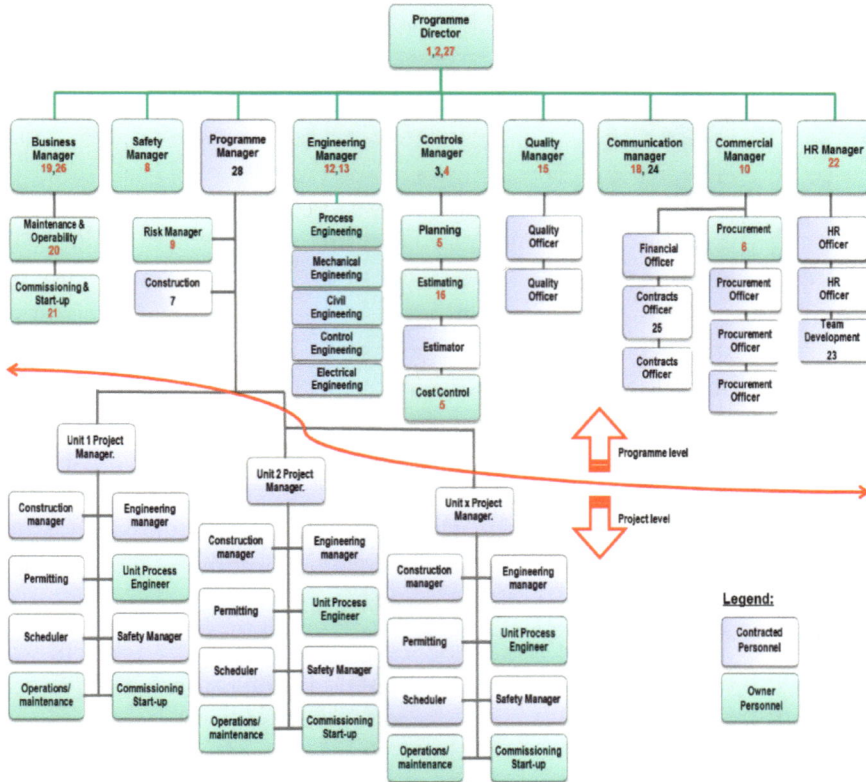

Figure 6.6: Integrated owner team concept

Key to the success of the programme, irrespective of which approach is followed, is not to outsource to contractor personnel those competencies and accountabilities identified as core. The absolute minimum requirement is to have owner personnel in the lead positions for all critical areas.

Now we turn to the matter of the physical programme management office, the layout thereof and the infrastructure required.

Establishing the programme office and infrastructure

Introduction

Because of the size, complexity and duration of a programme it is normally required to set up a programme office, from where the activities can be overseen. The programme office should cater for the needs of the team members, as well as the necessary infrastructure to facilitate all the necessary office interactions such as discussions, document handling, communication and secretarial services.

There is often severe resistance to setting up a programme office with key personnel together in the same location, specifically from management and personnel from the owner company. The benefits of locating all activities together far outweigh the initial resistance to such a move, because of the overall coordination and integrated support to be given to the individual projects comprising the programme.

It has long been known that when people work more than 30 meters apart, the likelihood that they will collaborate reduces dramatically. Within 30 meters, it is so much easier to be aware of the presence and activities of co-workers, and to initiate informal interactions as needed by the flow of the work. Thus, co-locating a work group in the same area is an especially productive arrangement. According to Olsen and Olsen (2003), collaborative work at a distance will be difficult to do for a long time, if not forever. There will always be advantages to being together, but as a wide range of collaborative tools emerges, it will be possible to use them to accomplish your goals. However, for megaprojects and programmes, we maintain that it is essential to set up a programme office.

Research suggests that the presence of others in the immediate surrounding increases attention, social impact, and familiarity (Kiesler & Cummings, 2002). These effects imply support for the saying, "out of sight, out of mind," with several implications for distributed work. That is, distributed work that causes people to be out of one another's sight may lead also to their comparative inattention to co-workers, to a lower level of effort, or to an increase in free riding. Face-to-face discussion has a strong impact on cooperation in work groups through its effects on bonds, social contracts and group identity, and it is the most powerful medium known for coordinating work within an interdependent group (Kiesler & Cummings, 2002).

The above mentioned points serve to illustrate the need for setting up a dedicated and integrated programme office. The issues, problems, discussions and solutions are unique to each programme or project. The fact that the work to be performed is by no means repetitive in nature requires frequent interaction between the team members.

It normally becomes clear very soon after establishing a programme office that real benefits are gained in locating the core team (programme staff (e.g. the programme director and support staff)) together. Various projects or parts of project lifecycles, like detail design work, will be executed in locations around the world. This implies that any programme will also have a 'virtual office' component in order to interact with all the different activities happening on the programme. However, the core leadership team should be able to work together effectively.

Essential features of a programme office

With the modern age of technology, communication issues are more easily solved through email, internet access, remote network dial-up software, mobile phones, laptop technologies and hand-held devices.

Specific software applications are also readily available to enable team members to interact with the project and programme databases, review and approve documents, load progress reports, etc., using smart phones and tablets. The programme office thus also contains the communications infrastructure and information technologies required to support the work.

A successful programme office environment will comprise the following components:

• A suitable location (easy access, both physically and for interconnectivity, access to overnight facilities for visitors);

• An office complex that is conducive to work i.e. sufficiently large, quiet, air conditioned and well illuminated;

• An office layout suitable for easy interaction between the programme team members with sufficient work spaces, formal conference rooms, informal meeting spaces, refreshment areas;

• Office Infrastructure:

    ○ Communications (telephones, telephone and video conferencing, e-mail, intra– and internet access, bulletin/notice boards, webcam facilities especially with other sites, project specific communication web portals, access to all other parties' document management systems);

    ○ Document handling (methodology, processes, distribution and approval, forms and registers, file storage, database storage and backup facilities);

• Programme specific tools (access to IT systems of the owner company and tools, procedures and services like contract management, financial management, project management systems, and risk management), and;

- Health, safety and environmental considerations, which are discussed below.

Office location

A programme will extend over a much longer period than a normal single project and team members will be expected to commit to this long duration and its potential impact on their personal lives. Deciding on the location of the programme office requires careful consideration, since it will almost certainly have repercussions. Care must be taken to minimise the impact of the programme on key personnel and their families as much as possible. The intent should be to minimise the misery in terms of relocation or long distances to be covered. When establishing a programme office, three options exist:

- Key team members have to physically relocate to the location of the programme office;

- Programme office location is accessible by public or private transport on a daily basis, and;

- Remote team members can have facilities where they can stay over during the week and return home during the week-end.

The safety aspects also need careful consideration. Will the chosen site for the programme office be in a safe and secure area, will it involve regular travel along dangerous routes and will long distances have to be travelled by the team members? In addition, the specific office building's adherence to the local regulations in terms of safety, escape routes, fire detection and fighting and such aspect needs careful review.

Sometimes the desired location for a programme office will be obvious and other times it may be more of a problem. Careful planning when siting a programme office will prevent the organisation from losing key programme personnel due to safety incidents, accidents and family pressure.

Office layout

For a work–based team such as a programme management team, employees do much of their work in a team environment. The preferred office layout for this way of working is an open environment that facilitates communication and a large number of small conference areas.

Consideration for an effective office layout can include an open bullpen design with some common areas in the centre. Several meeting rooms should be provided which are easily accessible; some to be reserved, others on a first-come, first-served basis. The meeting rooms can also double as idea rooms and must be equipped with flip charts, whiteboards and conferencing facilities (e.g. Windows LiveMeeting, Skype or similar set-ups). Consideration should be given to walls, to be able to stick charts and reports on, printer access, lots of pens and table space. The arrangement should ultimately be based around the specific features or requirements of the programme. Daily stand-up meetings can happen in the middle of the open-space work area, resulting in effective communication.

Senior members of the team often want privacy to conduct work discussions with subordinates and therefore the open area can be surrounded by the managerial offices.

Figure 6.7 is an example of a programme office layout that should encourage overall programme interaction and integration. Remember to allow at least 4,5m^2 per member of the team for his or her dedicated workspace. The use of common resources across the programme enables optimisation of resources and management of the issues pertinent to a programme as mentioned before.

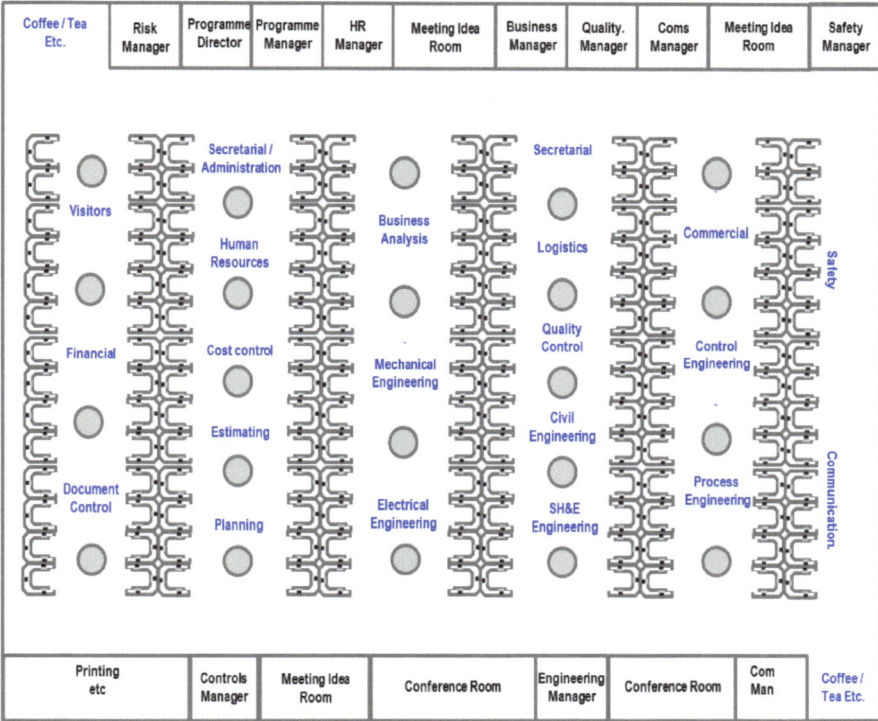

Coffee / Tea Etc.	Risk Manager	Programme Director	Programme Manager	HR Manager	Meeting Idea Room	Business Manager	Quality. Manager	Coms Manager	Meeting Idea Room	Safety Manager

Visitors — Secretarial / Administration — Business Analysis — Secretarial — Commercial

Financial — Human Resources — Logistics

Document Control — Cost control — Mechanical Engineering — Quality Control — Control Engineering

Estimating — Civil Engineering — Process Engineering

Planning — Electrical Engineering — SH&E Engineering

Safety — Communication

Printing etc		Controls Manager	Meeting Idea Room	Conference Room	Engineering Manager	Conference Room	Com Man	Coffee / Tea Etc.

Figure 6.7: Illustrative programme management office layout

The specific sub-projects also need dedicated resources over and above the common support resources allowed for in Figure 6.7. Each project may need a dedicated project manager, engineering manager, process engineer, operations, maintenance and commissioning personnel and a cost controller. Initially these project teams may have offices adjacent to the programme team, but some of the personnel may move to engineering contractor's offices during the design and procurement phases and to the site where the project will be constructed after that.

153

Office infrastructure

The programme office will have to handle large volumes of electronic data and need to be set up to ease communication between different parties that may be in various locations globally.

Firstly high speed broadband internet access is essential and must be set up from the start. Document handling needs to include an electronic document management system where documents can be stored, development progress tracked, electronically reviewed, approved, transmitted to other parties and printed. It may also require access to the document handling systems of other project teams.

Concluding Remarks

Taking due care while shaping the programme organisation will ensure a properly staffed structure with the correct competencies, personnel that understand their roles and responsibilities and a programme management office that functions effectively.

For an owner organisation, setting up a properly staffed and equipped programme management office is very expensive in terms of salaries and office space requirements. However, for a programme or megaproject to be successful, the owner organisation cannot skimp on these expenditures.

Chapter 7:
Maintaining Control of the Programme

"One of the true tests of leadership is the ability to recognize a problem before it becomes an emergency".
- Arnold Glasow

Introduction

The execution of a programme and associated projects is based on a set of robust plans and success can only be achieved through an effective control methodology for quality, scope, cost, and schedule. Therefore control is the fourth element or the fourth blade of the propeller in our programme management model, as shown in Figure 7.1. It is widely recognised that ineffective planning and monitoring plays a major role as the cause of project failures.

The development of a suitable control system is an important part of the programme management effort. The reason is simple: good control reveals problems early and that means you'll have more time to take corrective action.

Project controls are the data gathering, management and analytical processes used to predict, understand and constructively influence the time, cost and quality outcomes of a project or programme; through the communication of information in regular intervals and in formats that assist effective management and decision making (adapted from Weaver, 2013). This definition encompasses all stages of a project or programme's lifecycle from the initial estimating needed to 'size' a proposed project, through to the forensic analysis needed to understand the causes of failure.

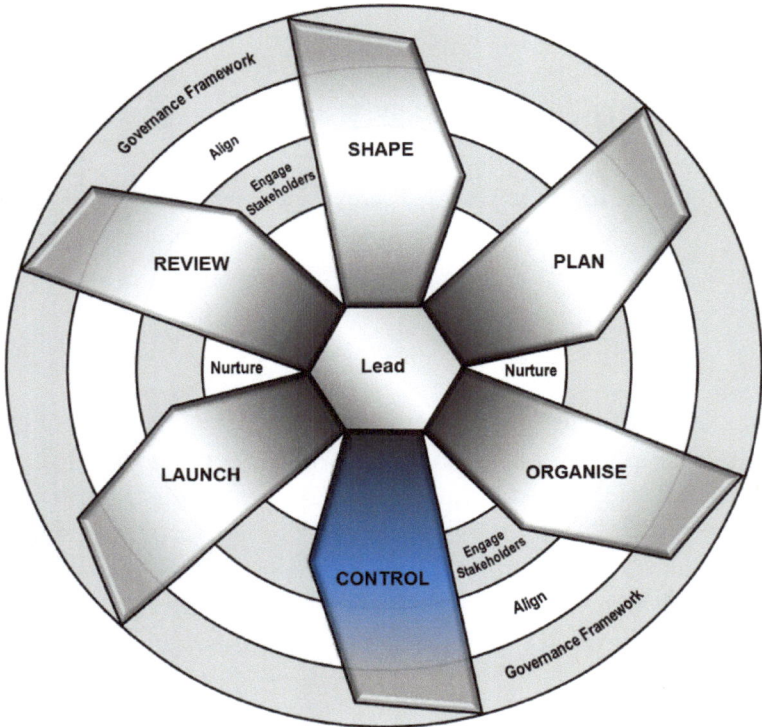

Figure 7.1: Programme management model with 'Control' highlighted

Control on a programme refers to four main elements, namely:

- **Quality:** Are all deliverables subjected to quality assurance protocols? Will the final outcome be performance as expected?

- **Scope:** Has the scope of the programme been fixed or is scope creep on the sub-projects still rampant?

- **Cost:** How does the actual expenditure compare to the forecast expenditure? Will the total programme cost overrun the budget?

- **Schedule:** Are we meeting the milestone dates according to the programme schedule? Will the projects be completed in time to achieve the required synergies?

Each of these elements is discussed in this chapter. The importance of having rigorous control processes for all four of these elements is explored.

Quality assurance and control

Introduction

According to Aristotle, "Quality is not an act. It is a habit". This implies that quality and quality assurance is a way of life, which includes the work situation. Quality cannot be inspected in, it has to be designed and built right.

As a first point of departure, quality assurance aspects on programmes are discussed. Quality is often referred to as 'fit for purpose' quality, implying that if a unit will only be required to operate for a year, it will not be manufactured to the same standard as a unit that is expected to operate efficiently for 30 years. Quality assurance and quality control, however, refer to ensuring that what is delivered meets the agreed project or programme quality requirements. Quality assurance and quality control thus includes the quality of the business plan, the engineering work in developing the programme specification, the detailed engineering design work, the manufacturing of equipment and construction of the hardware, up to finally running the facilities as an on-going business venture. This also implies quality of project execution planning.

Quality assurance philosophy

Quality assurance or control is of critical importance to ensure a programme meets schedule, is within budget (the programme budget and sub-project budgets) and that meets the overall business objectives. Rework is often required once quality of the workmanship deteriorates. This inevitably leads to cost, schedule, scope or ultimately operational impacts. It can almost be said that if you look after the quality, the other aspects (i.e. cost and schedule) will look after themselves. Certainly, if quality is not ensured, it will be almost impossible to meet any of the other objectives.

On a programme, a large percentage of the quality assurance work is typically carried out by common programme resources. It is therefore recommended that a programme wide quality management philosophy is developed and enforced. Each sub-project management team must still implement a project specific quality management plan, conforming to the programme quality philosophy and guidelines.

In addition to the project quality plans, certain activities are grouped together and conducted or managed centrally by the programme team. The programme management team therefore needs to develop their own specific quality plans that covers all activities performed by the programme management team themselves. These activities include the auditing and monitoring activities that the programme team may be performing to ensure adherence to the agreed quality plans.

The programme management quality assurance philosophy and guidelines should explain the quality assurance concepts as they will be applied, as well as the accountabilities and responsibilities of the different parties or teams. Typical elements to be covered in the programme quality assurance philosophy are:

- **Quality Management Strategy:** The overall quality management structure, accountabilities, responsibilities and any specific overall strategic guidelines;

- **Quality Tactics:** The tactical requirements for

implementation of the quality management plan. Highlight overall guidelines in terms of hold, review and approval requirements. These include stage gate requirements as well as pre-defined interim checks;

- **Reporting Requirements:** This covers the requirements, formats, templates and frequency of reporting quality assurance activities, deviations and corrective actions. The purpose of the project quality reports is to inform the programme management team of the status and progress made on project quality management in order to ensure overall integrity, and;

- **Monitoring:** The programme management team has to audit, review and monitor the quality management performance of the different sub-project teams to ensure that quality plans are actually being implemented and followed through.

To illustrate how these four aspects are practically implemented, examples of the Intego quality strategy, tactics, reporting requirements and monitoring requirements will now follow. Let's start with the quality strategy.

Intego Environmental Compliance Programme (IECP)

Overall Programme Quality Assurance philosophy

Objective

The definition of reaching success from a quality perspective will be that the performer of an activity takes ownership of that activity and delivers to the agreed requirements, at the agreed, or better, cost and schedule targets.

Introduction & Background

The purpose of this document is to define the overall quality philosophy through which the programme owner team provides the requirements and monitors the quality of the Intego Environmental Compliance Programme. This will result in the responsibility for executing the actual work being delegated to the correct party, and ensures that quality management work is not duplicated and is performed at the correct level.

Based on the philosophies set out in this document, project specific quality management plans will be developed for each individual project within the programme, coordinated by the relevant project manager. Project specific quality plans shall comply in philosophy and execution with this document. Quality assurance activities will cover all work executed under the realm of the programme whether it is the development of business plans, the development of programme or project level specific project management, engineering or operations and maintenance specifications as well as engineering design, manufacturing, construction, commissioning and start-up activities.

Quality management strategy

General

The IECP is a multi-billion dollar venture comprising various projects at different locations. Because of the duration, magnitude and complexity of this programme, it is even more essential than usual to identify and agree the critical quality assurance (QA) activities that the programme management team will engage in. Also included is the quality assurance and control (QA & QC) activities that both the project teams, engineering contractors (ECs), suppliers, manufacturers and constructors need to perform.

The essential premise of the programme quality philosophy is that the responsible party who performs the work needs to demonstrate that they can meet the agreed standard and must ensure the quality of said work. The programme team should fundamentally never have to identify or uncover non-conformances or trends in poor quality issues as they need to be identified by the project and EC teams before they reach the programme team. This implies that the programme team, during the performance of the work, may identify individual quality issues or omissions; however trends or non-conformances should have been corrected by the relevant EC before the programme team personnel become involved.

When a quality non-conformance or trend in poor quality is identified in work being executed, the root cause needs to be identified and addressed by the executing party. If the programme team does identify such issues, then they will intervene. When a non-conformance is identified by the programme team, the intervention will include stopping specific work or activities associated with that part of the work. The executing party is then obligated to correct the non-conformance.

Quality strategy

The quality strategy for IECP is graphically represented in Figure 7.2. Based on Figure 7.2, the following comments can be made:

- A specific operating unit will finally accept control of the new plant and facilities. Therefore, any specific quality requirements of the operating unit need to be taken into account. Overall quality guidance and programme quality guidance will be approved by the programme sponsor on recommendation of the programme steering committee. The operating unit will thus perform QA activities during the course of the project as part of the planned project activities to confirm that the scope of supply is consistent with the charter and agreed project specific requirements;

161

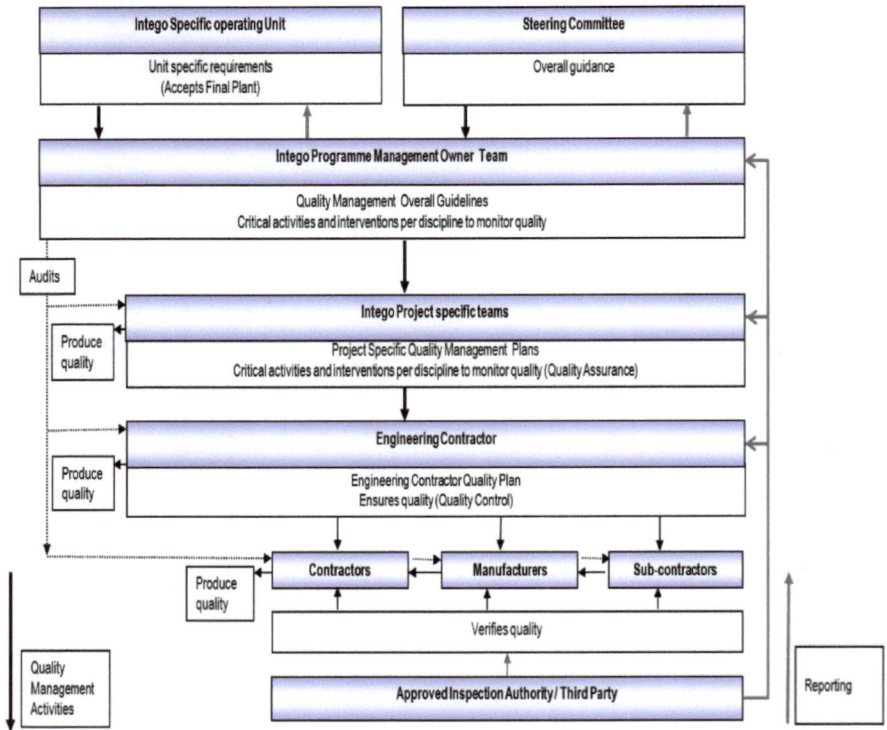

Figure 7.2: Quality management strategy

- The programme management team sets the overall guidelines and monitors implementation of the plans at all levels. Any general trends, common quality management issues and non-conformances will be highlighted and addressed by the programme management team. The programme management team will report on quality management issues to the programme steering committee;

- Each project team will develop and implement a detailed project specific quality management plan. Each project team will manage the agreed hold and review points and approve predetermined key documents. Common programme resources will be utilised for these reviews and approvals where a specific project team does not have dedicated resources for such activities, and;

• Each EC is depicted as responsible for their own quality and the quality of their subcontractors and manufacturers.

Quality Tactics

General

This section describes the reviews and checks required by each discipline to assure that deliverables and equipment are produced to the required quality standards.

Review attendance

A multitude of review steps are required throughout the programme and sub-projects' lifecycles. In this paragraph the reviews that need multi-discipline participation is addressed.

Table 7.1 depicts the representation that is required at the programme and project reviews. The attendance is prescribed from a programme level to ensure consistency in the overall approach across the complete programme. The individuals with the highest loading by far include the engineering manager, the lead process, mechanical, electrical and control engineers, the SH&E representative and the process operations representative.

The table represents the essential participants in the different reviews. Specific circumstances may necessitate the attendance of other specialists as well.

Table 7.1: Review meeting attendance table

Intervention Point	Engineering Manager	SH&E Representative	Commissioning Manager	Lead Process Engineer	Lead Mechanical Engineer	Lead Welding Engineer	Lead Mechanical QA	Lead Rotating Engineer	Lead Electrical Engineer	Lead Control Engineer	Mechanical Maintenance	Control Maintenance	Electrical Maintenance	Rotating Equipment Maintenance	Process Operations	Lead Information Management
Framing & Alignment	Y		Y	Y	Y				Y	Y					Y	
Kick-off Meeting	Y	Y		Y	Y				Y						Y	Y
Business Case review	Y		Y												Y	
Business Assumptions Review	Y	Y	Y												Y	
Technology & Licensor Selection	Y		Y												Y	
Specification Selection	Y				Y	Y	Y	Y	Y	Y	Y	Y	Y	Y	Y	
PFD Review	Y	Y	Y	Y	Y					Y	Y	Y			Y	
Site Flowsheet & Integration Review	Y			Y					Y	Y					Y	
Design Basis Review				Y						Y					Y	
Conceptual Engineering Package Review	Y	Y	Y	Y	Y				Y	Y			Y			
Gate Readiness Review– Engineering	Y	Y		Y	Y				Y	Y						Y
Business Framing & Alignment	Y			Y	Y					Y			Y			Y
Kick-off (MC with EC)	Y	Y		Y	Y					Y					Y	Y
Process Design Basis				Y						Y						
Draft Programme Vendor List	Y			Y	Y				Y	Y	Y	Y	Y			
Materials of Construction Review	Y			Y	Y		Y				Y					
Preliminary Identification of Long-lead items	Y			Y	Y			Y	Y	Y						
Process & Instrumentation Diagram Review	Y	Y	Y	Y	Y	Y	Y	Y	Y	Y	Y	Y	Y	Y	Y	Y
Finalised Long-lead items list	Y			Y	Y			Y	Y	Y						
HAZOP and Safety Integrity Level Review	Y	Y	Y	Y	Y	Y	Y	Y	Y	Y	Y	Y	Y	Y	Y	Y
Review and sign-off Mechanical Datasheets	Y				Y	Y	Y	Y			Y					
Model Reviews	Y		Y	Y	Y					Y		Y	Y	Y	Y	
Basic Engineering Package Review	Y	Y	Y	Y	Y	Y	Y	Y	Y	Y	Y	Y	Y	Y	Y	Y

Legend: Y = Yes

Project specific quality plans

For each project within the programme a quality plan covering each discipline (e.g. business development, operations & maintenance, commissioning & start-up, project management, engineering design, procurement, manufacturing and construction) will be developed for all activities conducted under the direction of the specific project team. The quality plan needs to be specifically developed for each project lifecycle stage (project definition, design development, execution and launch) that it has to be applied to. The template as provided in Table 7.2 is to be used for the project specific quality plans.

Table 7.2: Template for Discipline specific quality plans per project

IECP		Quality plan					
		FBS NUMBER (SYSTEM):					
PROJECT NAME:		Rev.	Page:	1	of	1	
DISCIPLINE:		Date:					
EQUIPMENT TYPE:		DISCIPLINE SPECIALIST:					

ITEMS/DETAILS TO BE APPRAISED	STATUS	Reviewed By	Criticality Rating & type of review

STATUS LEGEND
X = Review Requested
C = Review Completed
PC = Partially Completed

NOTES

The following principles pertain to the quality plan:

- Quality assurance will be done on a 90/10 basis. Critical deliverables, first of a kind deliverables and unique deliverables will be considered. Approximately 10% of these deliverables shall be subjected to quality assurance (QA) audits;

- The status of a review can only be marked as completed if the quality requirements have been met. All non-conformances need to be referred back for rectification, and;

- The criticality and thus the type of review required must be noted. For example, if an engineering design is to be reviewed, the review can range from a very superficial check for consistency only up to a detailed check of the design itself. A design check can be performed using a different design package or even a check based on fundamental principles.

Programme specific quality activities

The programme management team will be responsible for the following specific activities:

- Quality management planning and implementation on bulk materials including non-conformance tracking and mitigation;

- Coordination of third party inspections;

- Issuance and control of specifications;

- Analysis of quality reports from project teams, identification of common trends and mitigation, and;

- Reporting on overall quality assurance to the programme steering committee.

Reporting

General

The purpose of the project quality reports is to inform the programme management team of the status and progress made on project quality management in order to ensure overall programme integrity. The individual project reports roll up into a monthly programme quality report.

Monthly quality tracking report

Each project team will produce a consolidated quality report on a monthly basis. The format of the report shall be as outlined in Table 7.3. Traffic light signalling is used to indicate the status of the different sections of the report, with a red light indicating a problem, an orange light a concern and a green light indicates that everything is on track.

Table 7.3: Format of the monthly project quality report

Quality Assurance Progress Report: Project:
Date: Project Manager:
1.   **Executive Summary**

2.	Discipline Key Activities/Concerns/Highlights/Mitigation	
2.1 Business Development		
Activity	Comments	Status 🟢
2.2 Project Management		
Activity	Comments	Status 🟢
2.3 Engineering Design		
Activity	Comments	Status 🟢
2.4 Procurement		
Activity	Comments	Status 🟢
2.5 Manufacturing		
Activity	Comments	Status 🟠

2.6	Construction	
Activity	Comments	Status
		🔴

2.7	Commissioning	
Activity	Comments	Status
		🟢

2.8	Start-up	
Activity	Comments	Status
		🟠

2.9	Performance Tests	
Activity	Comments	Status
		🔴

Legend: 🔴 Problem encountered 🟠 Concern identified 🟢 Everything according to plan

The status indicators shown in Table 7.3 are only as an example. The choice of a red, orange or green traffic light will obviously depend on the actual circumstances pertaining to that component of the report. These status indicators should be reviewed every time the monthly quality report is updated.

169

Monitoring

The programme management team will:

- Review and approve project quality management plans;

- Receive and perform quality checks of the monthly project quality reports;

- Consolidate the quality reports into an overall quality report for the programme, and;

- Conduct systems and technical audits to ensure that the projects are executed according to the quality plans.

Document control for quality

It is normal practice in the engineering disciplines to address quality and especially quality in manufacturing. However, this is not enough to ensure an overall quality project. Business planning and development, operations and maintenance and project management activities should also be subjected to quality management.

The first quality management principle to implement is document control. Documents should indicate who the author, the reviewer and the final approver are. These should be minimum requirements. Too often documents are generated with no review or approval cycle. Mistakes or oversights only become apparent too late; resulting in costly rework or work-around plans.

Quality tracking report

The template for the quality tracking report has been discussed in the overall programme quality assurance philosophy for the Intego case study in the preceding pages.

Intego Environmental Compliance Programme (IECP)

Quality Tracking Report

Date: End of Month 6

Executive summary

The overall programme economic model was reviewed by an independent auditor and no errors or inconsistencies were found.

Audits were conducted at four of the engineering contractors and it was found that the engineering design documents were not consistently checked and approved as per their own internal procedures. Errors were found on some of the datasheets reviewed. This specifically pertained to the mechanical and electrical disciplines.

Non-adherence to quality procedures were found in the procurement departments of the engineering contractors. Technical proposals from bids were not reviewed and signed-off by the relevant engineering disciplines in all cases. Tenders were found to be ready for award with the actual technical information not conforming to the requirements in the enquiry.

The schedule produced for Gas 1 NOx removal was reviewed and no errors were found in the assumptions, critical path analysis and resource loading.

The quality status summary report is presented as Table 7.4. Note that this is only an extract of the complete summary report and does not include all disciplines. The objective is to show how easily one can pinpoint problem areas.

Table 7.4: Extract from quality status summary

Activity		Quality Monitoring			
		Mechanical Engineering	Electrical Engineering	Instr. & Control Engineering	Civil Engineering
Engineering	Quantity	🟠	🟠	🟢	🟢
Engineering	Quality	🟠	🟠	🟢	🟢
Procurement	Quantity	🟠	🟠	🟠	🟢
Procurement	Quality	🟠	🟠	🟠	🟢

Legend: 🔴 Problem encountered   🟠 Concern identified   🟢 Everything according to plan

Mitigating actions

Non-conformance reports were issued to the relevant contractors. A re-audit will be conducted within 30 days. Those contractors, who fail to take corrective action in the time allowed, will be dealt with according to the penalty clauses in their respective contracts.

Scope Control

Introduction

Fit-for-purpose quality control on a programme is clearly of utmost importance. The next major control area for discussion is control of the scope of the programme and projects.

An anonymous quotation states that anything that can be changed will be changed until there is no time left to change anything. On a megaproject or major programme stretching over many years, changes in key role players are inevitable. Scope management is therefore much more important on programmes than on single large projects. Analysis has shown that the majority of large complex projects inevitably suffer from scope management inadequacies.

Veryard (2001) states that the opportunity for scope-creep is greatly enhanced for a large programme which consists of many sub-projects. He says that each project team has a vision of its scope, and perhaps a vision for the whole programme from its own viewpoint, leading each team to draw conclusions about the 'proper' place for this or that material. Even when two neighbouring projects agree about the boundary between them, a programme engineer with a different perspective over the whole programme may wish to overrule them.

It is the interfaces between the projects that often cause the greatest trouble. Here again, scope creep is at work, encouraging the project managers to creep across the project boundaries into the next project. This phenomenon might seem to give project managers some incentive to manage these risks collaboratively.

Programme management is therefore sometimes called upon to play a policing role: patrolling the boundaries to make sure the design team's vision is respected, and is not undermined or corrupted by scope-trading (Veryard, 2001).

The opportunity for scope-creep is greatly enhanced for a large programme which consists of many sub-projects (Veryard, 2001).

173

Managing scope changes

In Chapter 5 which focused on planning, the development of the facilities breakdown structure (FBS) (often also called the work breakdown structure) and work package scope definition was described. No planning or control can be done without an agreed scope.

The first revision of the scope definition package should be completed toward the end of the programme shaping phase or at the latest very early in the planning phase as indicated in Figure 7.3. Early definition of the scope helps to understand the scope and the cost and schedule implications thereof. It should also serve as the basis from which to track scope changes as the work becomes more defined.

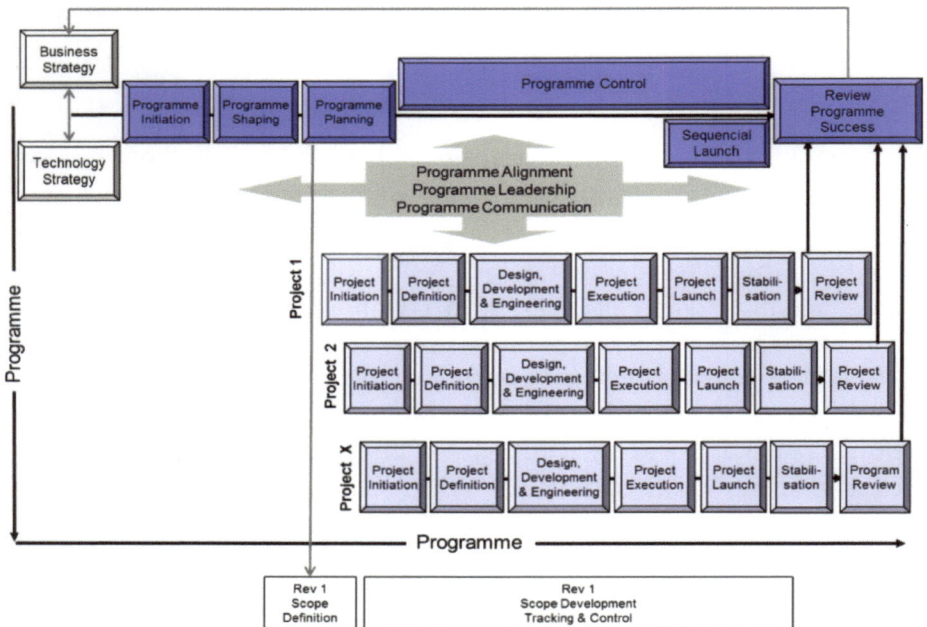

Figure 7.3: Scope definition and tracking

Initially, scope tracking could be a simple exercise where changes are noted and the implications in terms of cost, schedule and risk indicated in the relevant estimates and documents. As individual project definitions are completed, much more rigorous scope management must be implemented for those projects. The focus also includes how scope change on one project impacts the other in terms of risk, scope, schedule and cost. At the end of the project definition phase of a specific project, no changes should be allowed unless the design is proven to be unsafe or it is shown that the process will not work as designed.

No changes should be allowed at the end of the project definition phase of a specific project; unless the design is proven to be unsafe or it is shown that the process will not work as designed.

Scope change approval should not be delegated to the individual project teams, but should be retained at programme level. The project team members should submit a change request to the programme management team for consideration. A best practice is to have regular (weekly) or even additional ad hoc scope management discussions such that the project work can proceed without delay. Representation at these discussions must be such that a decision can be made during the discussion.

Scope change We return to the Intego case study to illustrate a scope change request. Note specifically the provision for signatures on the scope change request. These approvals and signing authorities must be established beforehand (refer to Chapter 14 on governance).

Intego Environmental Compliance Programme (IECP)

Scope change request

Project:	Coal 3 Environmental Compliance Project
Work package	Utilities
WBS Code	1.2.3.4
Change request Number	OOOO1
Date	1 June 2011
Recommended by Project Manager	Signature
Approved by Programme Management	Signature

Description of change

The change involves the inclusion of a common nitrogen production unit, consisting of membrane air separation, nitrogen compression and distribution.

Justification

It was envisaged in the original concept that nitrogen requirements will be low. It was considered most effective if a unit that requires nitrogen provides its own liquid nitrogen storage facility and purchases liquid nitrogen directly from a vendor.

It has since been determined that the nitrogen demand is such that providing a common membrane air separation unit with compression and distribution makes more business sense. At a capital outlay of US$ 10 million and an operating cost of US$ 0,3 million per annum, the project will have a payback of less than 2 years as compared to purchasing liquid nitrogen as per the present concept.

Capital cost impact

Central unit plus reticulation system:	US$ 10 million
Less saving on individual units:	US$ 2 million
Net capex:	US$ 8 million

Operating cost impact

Annual operating and maintenance cost:	US$ 0,3 million
Annual purchase cost of liquid nitrogen:	US$ 4 million

Schedule impact

The proposed membrane separators are standard off-the-shelf units. Delivery time for the compressor is 12 months. As the project is in the front-end-loading phase, it will not result in a delay in the anticipated schedule.

_____

Once the scope change is approved, all affected scope sheets as well as the facility breakdown structure (if required) should be updated to show clearly what has been removed or added to the different packages as well as the revised capital cost of each package, the sub-project and eventually the complete programme. All the affected sheets must be re-approved and the cost estimates and schedule updated to include the change.

A register with the latest revision numbers for all packages as well as the WBS must be kept up to date and accessible.

For engineers the above sounds like an unnecessary burden. However, rigour in this process will pay off in the long term. The implications of changes will be known early on and mitigating plans can be put in place to ensure that the programme stays on course and within the original targets. At least the overall implications can be discussed at senior level and either agreed to or the programme can even be terminated if no longer justifiable. Effective scope management will reduce sudden surprises and expensive rework at the end of phases where value engineering will be required to try and rescue the project. Scope control also avoids having to figure out why the cost estimate for the programme has increased at the end of a phase.

It is clear that scope control on a programme is critical. And if done well, will have a major positive impact on the programme success. The next area of control that deserves attention is cost control.

Effective scope control for a programme and its sub-projects is a prerequisite for programme success. Scope change approval should not be delegated to the individual project teams, but should be retained at programme level.

Cost control

Introduction

The cost control focus on a programme is not only on the control of an approved budget for a specific phase, but also on looking forward and tracking the predicted or forecast end-of-job cost from a very early stage. The development of any project or programme is primarily to satisfy a business need and not to develop a superb engineering package disregarding the business objective. As such, it is required from inception of the programme to have an idea of the overall cost. This of course depends on the scope and therefor the requirement to develop a first draft of the scope and an estimate of the cost of each work package as outlined in chapter 3. This scope definition and estimate is of a very rough order of magnitude nature and needs to include an appropriate contingency to take into account possible inaccuracies and uncertainties.

The principle is then to at least do a monthly update of the overall programme cost estimate, based on the information received from each project. This provides a running view of where each project, and eventually the total programme, is heading in terms the end-of-job cost. The programme team should have a clear understanding of the business objectives and the boundaries within which they have to manage the programme. If it appears that the cost is escalating more than can be accommodated, the specific area causing the escalation needs to be identified and appropriate actions implemented to correct the trend. This could include stopping the specific project and re-scoping it, finding alternative technology or maybe reducing the cost in another area to protect the overall programme viability.

Risk factors for project cost

Assuming that a thorough and accurate cost estimate is done for a project, the estimated cost is mainly affected by two causes, namely:

- **Scope of facility:** As the definition of a project is improved, it may become apparent that an additional processing step may be required, or a certain work package will require more pieces of equipment or that the design specification is more onerous and thus the cost of a piece of equipment rises. All these types of changes are included under project scope changes. The engineers, who should be familiar with the scope definition, must bring any significant changes to the attention of the management team, and;

- **External market conditions:** The second type of cost deviation is caused by changes in market conditions or perhaps selection of an alternative supplier, resulting in a change of the price of a specific piece of equipment, changes in labour rates, changes in the price of steel, etc. It is the responsibility of the cost engineers to keep ahead of these types of changes in the external environment and regularly update the cost base to indicate these so-called trends.

The final cost projection is therefore a combination of the effects of scope changes, as well as market trends.

End-of-job cost tracking

The programme team must provide guidelines to the project teams on how to report on scope changes and market trends. Cost impacts can then easily be consolidated for the complete programme. The result of this consolidated cost tracking exercise must be reported to the programme steering committee, which should include recommended mitigating actions to minimise the impact on the programme.

An example of a cost tracking report is included below for the Intego case study.

Intego Environmental Compliance Programme (IECP)

End-of-Job Cost Tracking Report

Date: End of Month 6

Executive summary

The end-of-job forecast has increased from $ 890 million to $ 930 million. Scope changes contributed $ 20 million and trends $ 10 million to the increase.

The additional nitrogen production unit contributed $ 8 million of the increase with a payback period of less than 2 years. It will result in an improved overall IRR and has been approved for inclusion.

An increase of $ 12 million has been caused by additional equipment for water clarification at Coal 1 as it was found that the water quality is actually worse than originally assumed. Additional sand filters have to be installed to accommodate the poor water quality.

During the reporting period the view on future steel price increases has resulted in a trend forecast increase of $ 10 million. This is due to a greater global demand for steel.

The development of programme cost against time is shown in Figure 7.4.

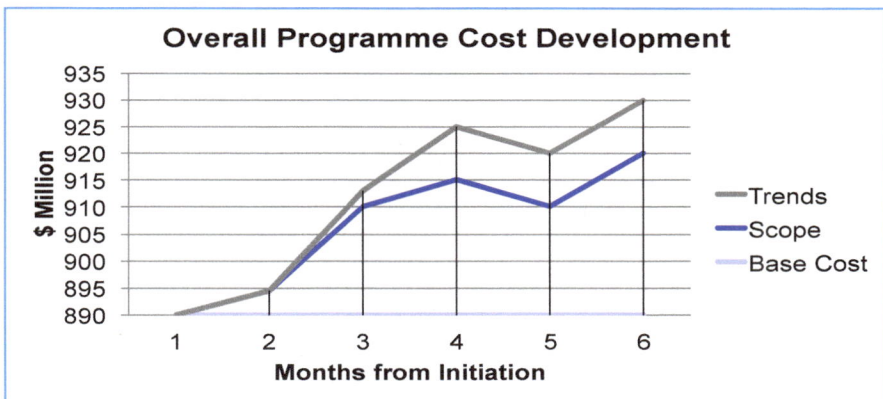

Figure 7.4: Cost development against time

Mitigating actions

A value engineering exercise will be conducted to optimise the water purification system for Coal 1, considering rapid clarification technology that may eliminate the need for extra filters. Water purification experts will be invited to participate in the exercise.

Further steel price increases are expected and the procurement department has been requested to investigate fixing steel prices by placing early orders on steel mills or forward buying of bulk steel.

The members of the steering committee may have additional insights or line of sight and may thus provide additional guidance.

It is clear that having a future view on the forecast end-of-job cost is critical on a programme to ensure the sustainability of the business venture.

Schedule control

Introduction

The fourth and last area of control to be explored is schedule control. Cost control is generally practiced to some extent on projects, but schedule control is more often neglected.

Scope changes may result in delays and additional construction labour. Inclement weather may result in delays and underestimation of the construction man-hours required. Either of the two will result in a schedule impact. This could be exacerbated by lower than required quantity estimates or lower labour efficiencies. Very little forward projection is normally done, resulting in an eventual scenario where there is much uncertainty in the schedule. Lack of schedule projection results in a syndrome of 'the project will be completed when it is completed'.

Master schedule tracking report

Various techniques and metrics exist for effective schedule control. The techniques include:

- Earned value techniques;

- Cost and schedule performance ratios;

- Reviews of scope;

- Quantity take-offs, and;

- Actual quantity progress measurements.

It is beyond the scope of this book to discuss these elements in detail, but again, as with scope and cost control, each project team should be given guidance as to the programme schedule management and reporting analysis requirements.

Reporting to the steering committee should be done by means of a master schedule tracking report. The tracking report should give the original schedule as well and the newer projections, deviations and mitigation actions and plans to protect the schedule. In order to be able to explain any deviations to the schedule, a schedule change register needs to be kept, in which all changes that affect the schedule are recorded.

Once again we return to the Intego case study to demonstrate what a master schedule tracking report should look like.

Intego Environmental Compliance Programme (IECP)

Master Schedule Tracking Report

Date: End of Month 6

Executive summary

Progress on Gas 1 and Gas 2 modifications have been slower than anticipated with both projects' schedule forecasts being around 6 months late. This has been caused by a scope change approved on the demineralised water feed systems to allow for the site requests for additional nozzles for shutdown purposes.

The Gas 3 forecast indicates that this project is approximately 3 months ahead of schedule as a result of faster ground preparation than planned.

On Coal 3 the schedule performance index is very low indicating poor productivity. The resultant expectation is a 15 month late completion, mainly on the particulate removal system.

The pertinent schedule impacts for the programme and sub-projects are reflected in Figure 7.5; master schedule tracking for phase 1 of the programme. Figure 7.5 shows the forecast and baseline schedule for each project. An area where actual performance is behind schedule is shown in red and where performance is ahead of schedule is shown in green.

Plans to mitigate the impact of schedule slippage have been developed and are presented below.

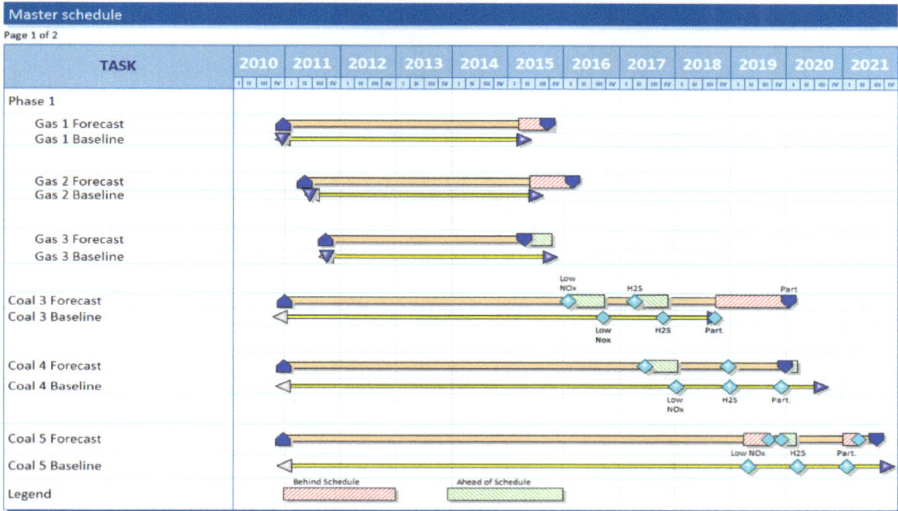

Figure 7.5: Master schedule tracking

Mitigating actions

On Gas 1 and Gas 2, the detailed schedules will be reviewed to establish the actual cause of the predicted delay and activities will be adjusted to reduce the delay.

The cause of poor productivity on Coal 3 will be subjected to a root-cause analysis and steps will be implemented to address the causes identified.

Concluding Remarks

The essential issue is to summarise and present the consolidated information in such a manner that the programme leadership and steering committee have a transparent view of how the overall programme is progressing. Consolidated information shows whether the programme is still on track to deliver the overall objectives as agreed and presented to the approval bodies. If this is done in a pro-active way, corrective actions can be taken where it deviates from the original mandate. The stakeholders can be informed about outcomes that may have changed, but are still overall acceptable without them having to face sudden surprises at a time when it is too late to attempt any fixes.

In summary, when it comes to control of quality, scope, cost and schedule, these activities take on a different nature at the programme level. Detailed control still needs to be executed on project level. To know exactly how your programme is performing, an integrated overview is required. In support of having a transparent view of quality, scope, cost and schedule and the risks associated therewith, it is best practice to have an integrated risk review session involving all key programme and project leaders.

Chapter 8:
Sequential Launching of Project Outcomes

"[Bill] Gates is the ultimate programming machine. He believes everything can be defined, examined, reduced to essentials, and rearranged into a logical sequence that will achieve a particular goal." - Stewart Alsop

Introduction

Launch, or the sequential launching of the outcomes of projects in a programme, is the fifth step in our programme management model, as depicted in Figure 8.1

Launching refers to the commissioning and start-up phase of programmes. The launching of individual projects is not the topic of this chapter as that is dealt with extensively in project management literature. The focus here is mainly on how to keep track of the overall business objectives or benefits of a programme. Sequential launching of project outcomes is how we ensure that the business objectives are met.

During the execution of a programme, the sub-projects are normally completed sequentially over an extended time period. During the planning and execution of the programme, steps must be included such that it is possible to measure and compare the eventual achievements against the objectives originally set.

In this chapter we firstly discuss benefits management and then move on to sequential launching.

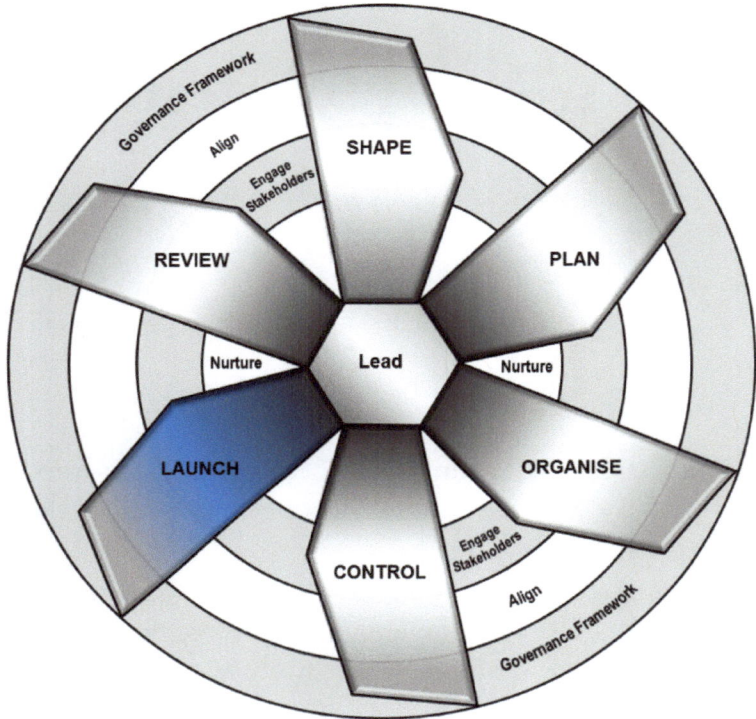

Figure 8.1: Programme management model with 'Launch' highlighted

Benefits management

Introduction

A great deal of time normally elapses between the time that a programme is started and its eventual completion. Things can change along the way as market conditions change, scope creep occurs and your key equipment deliveries are delayed. Small shifts in process design and project execution can affect the timing and quality of the benefits realised. Benefits management means that the desired business objectives and targeted benefits are fully satisfied at completion of the programme.

There is a risk that the benefits of a programme may be lost if there is a weak link between the programme deliverables and the business needs, or if the needs are not well understood by the programme management team. This risk is exacerbated if the programme team does not form part of the owner organisation. It is essential to establish a clear business case for the programme, so that you can make sure that the deliverables meet expectations and give the organisation the benefits it requires and expects.

There is a risk that the benefits of a programme may be lost if there is a weak link between the programme deliverables and the business needs. This risk is exacerbated if the programme team is separate from the owner organisation.

Business benefits are the reason any programme or project is created and implemented. According to a posting on Mindtools (2014), benefits management is all about ensuring that the hard work and investment that's gone into the project gives the greatest possible business return. It's important to focus on the programme's benefits, and not just its timeous completion.

Benefits management forces you to stay focused on why you started a programme in the first place. It doesn't stop after the programme ends, but continues until all benefits are clearly achieved. You can use the same planning framework as the rest of the programme to do this, but you'll need to build in benefit-specific milestones, as well as establishing clear accountabilities, and setting up appropriate communications systems. Done this way, benefits management can be a smart addition to a comprehensive programme management plan (Mindtools, 2014).

As compared to a project, tracking and being able to prove that benefits (business objectives) have been achieved on a programme is more challenging. Programmes are generally shaped and executed over an extended time period. In addition, the outcomes of various sub-projects may jointly contribute to a specific business objective. The overall benefit is only realised when these projects are completed, even though the completion date for the projects may differ.

Baseline performance

The first task is to establish the baseline performance of each of the business objectives before any changes are implemented. This has to be done by accurate measurement of the actual performance of the existing facilities at a specific point in time or over a predetermined period.

For the purposes of benefits management, the baseline performance of the parameters being addressed by the programme must be determined. Baseline performance can be measured for any of the business parameters, including:

- Production volumes of specific products;

- Product quality as determined from spot checks, customer complaints or returns;

- Sales volumes of specific products;

- Stock holding and turnover;

- Financial measures such as production cost, profit margin and labour cost;

- Unscheduled equipment down-time, and;

- Environmental footprint.

Environmental footprint is a collective name of the total environmental impacts of a facility, product or business. Although every business should be aware of all their emissions and impacts, the environmental loading in terms of the pollutants of concern i.e. greenhouse gases, SOx, NOx and airborne particulate material is of particular importance.

Once the baseline performance has been determined for the business parameters pertaining to the programme objectives, it should be signed-off by both the business unit and the programme management team. This agreed baseline performance is the measure against which any performance improvement as a result of the programme will be compared.

Establishing the baseline at the start of the programme development sets the design basis from which the improvements are to be made. The design and execution of the sub-projects can take a number of years to complete and by the time the project is launched the actual performance may not be anywhere near the original baseline because of potential deterioration in unit performance due to wear and tear, incorrect operating procedures and units operating beyond their design capacities.

Programme contribution

The programme contribution can be either positive or negative compared to the baseline, depending on the specific business objectives. This is depicted in Figure 8.2, where two bar-graphs are shown; the one on the left reflects a positive programme contribution and the one on the right reflects a negative contribution.

If the business objective is to improve production capacity or to increase profitability, the baseline at the start of the programme would be lower than the desired end-state. This means that the programme contribution is positive and increases the baseline to a higher level, as shown in the left hand bar-graph.

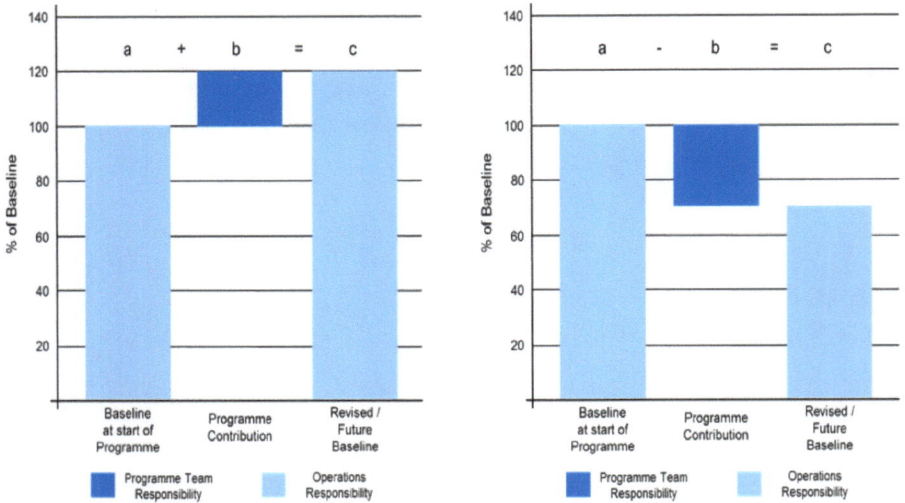

Figure 8.2: Baseline and programme contribution gives future state

Establishing baseline performance at the start of programme development sets the design basis from which the improvements are to be made. Depending on the parameters used, improvements can either be positive or negative.

In other cases, the business objective could be to reduce the baseline performance to a lower level. The Intego case study is one such an example where the desired end-state is to have lower SOx, NOx and particulate matter emissions. Another example is where the business objective is to reduce unplanned down-time at a specific facility by a certain percentage. In these cases the programme contribution is thus negative, as reflected in the bar-graph on the right hand side.

Future baseline

The future baseline is simply the reference baseline at the start of the programme and taking into account the programme contribution. Ideally it should represent the business objectives for the specific parameters addressed by the programme.

In the case of an unsuccessful programme or project, the future baseline may be different from the desired end state and business objectives. This implies that further work will have to be done to achieve the business objectives. In some cases the future baseline may be significantly better than the business objectives, which could indicate unnecessary expenditure on the programme, or that a more effective technology was used than originally foreseen.

The future baseline is the reference baseline at the start of the programme and taking into account the programme contribution. It should represent the business objectives for the specific parameters addressed by the programme.

In order to illustrate the concepts of baseline performance, programme contribution and desired end-state, we refer back to the Intego case study.

Intego Environmental Compliance Programme (IECP)

Baseline performance and desired end-state performance

Desired end-state

The emission reduction objectives as discussed for Intego Holdings in Chapter 4 are repeated here. The objective is to reduce SOx, NOx and particulate matter (PM-10) emissions by 10% in 2015 (Phase 1), 20% in 2020 (Phase 2) and 30% in 2025 (Phase 3) as compared to the 2010 baseline.

A major assumption is that government will opt for regulations lower that the business objective set in the charter, namely reduction of SOx, NOx and particulate matter by 10% in 2015, 20% in 2020 and 30% in 2025. A strategic decision was taken to take a progressive stance to also accommodate possible future changes in legislation.

The indication of potential emission reductions against a timeline is shown in Figure 8.3, which shows the downward trend for each of the three pollutants of interest (SOx, NOx and particulates) from the 2010 baseline; the baseline is indicated as 100% of emission load. Figure 8.3 shows a step change in performance after each of the three phases of the programme.

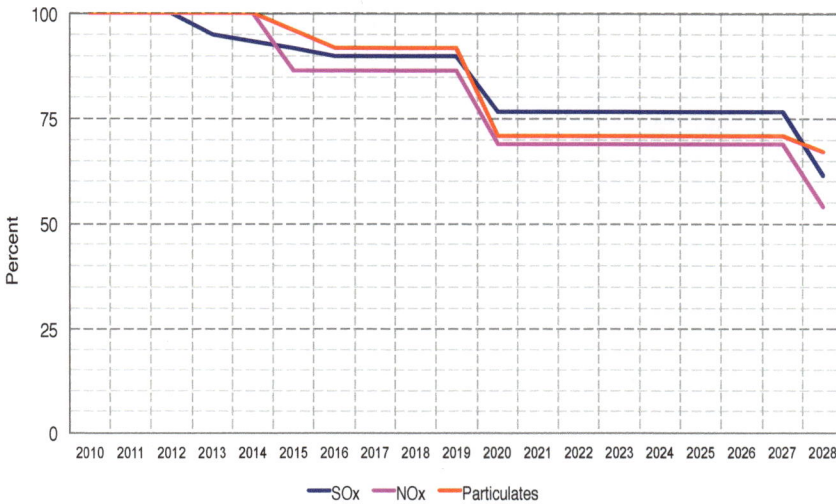

Figure 8.3: Desired end-state for pollutants of interest

Baseline performance

We will only concentrate on the NOx emissions for our discussion. The process will have to be repeated for each of the emissions of interest.

The NOx emissions from the different facilities of Intego Holdings are shown in Table 8.1. The total tons of NOx generated per annum are given for each facility, as well as the percentage of the total mass of NOx generated. The emissions reported in Table 8.1 were measured during the third quarter of 2010.

For the Intego case study, the baseline as measured and reported in Table 8.1 remains the legislative basis from which the improvement has to be made in absolute mass emission terms.

Table 8.1: Baseline NOx emissions as measured during 3rd quarter 2010.

Source	NOx Emissions	
	Tons/Annum	% of Total
Coal 1	663,75	25
Coal 2	531,00	20
Coal 3	265,50	10
Coal 4	265,50	10
Coal 5	265,50	10
Gas 1	265,50	10
Gas 2	132,75	5
Gas 3	132,75	5
Gas Production	79,65	3
Mining	53,10	2
Total	2655,0	100

The NOx emissions as reported in Table 8.1 have been re-measured quarterly and reported to the steering committee. The results for the period up to the end of Quarter 1 2015 are indicated in Figure 8.4.

Two baseline excursions were identified and rectified by the operations team that ensured the baseline remained intact. The reduction in NOx emissions from the launching of the abatement projects on Gas 1 and 2 has delivered the reduction objectives, but six months later than scheduled.

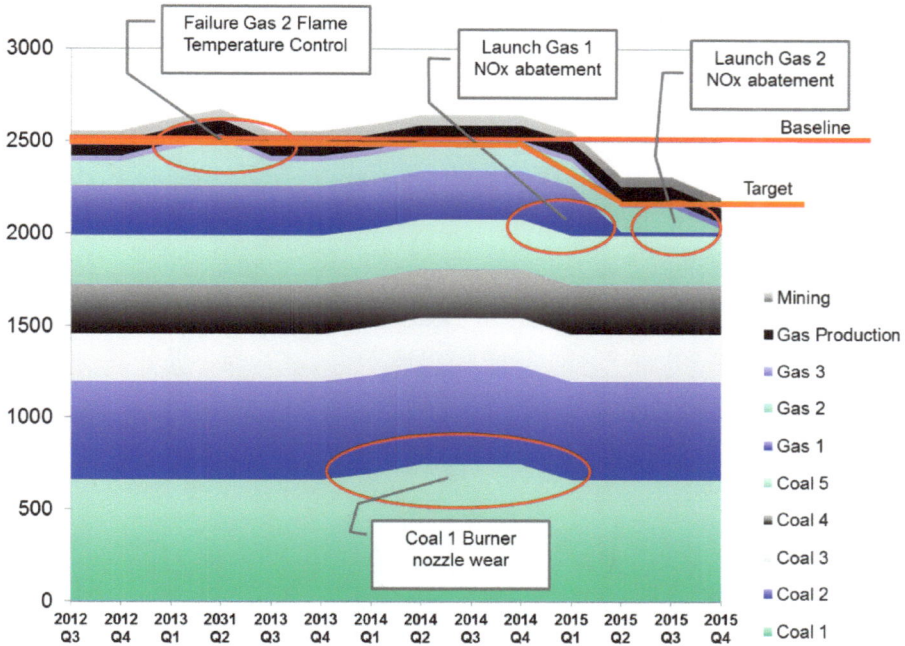

Figure 8.4: NOx Emissions Report

Regularly tracking the actual performance against the baseline as used in the design base is of utmost importance throughout the lifecycle of the programme and any deviation must be noted and actions agreed to mitigate the implication of any deviation.

Generally it is the responsibility of the operations department to report and maintain the performance of the operating units against the agreed baseline while the programme team is held responsible for the delta improvement.

197

If this concept is applied, it will always be clear as to whether the programme has actually achieved the improvement as set out.

It is of utmost importance throughout the lifecycle of a programme to regularly track actual performance against the baseline performance. Any deviation must be noted and actions agreed to mitigate the implication thereof.

Sequential launching

Introduction

According to the Oxford Dictionary, a sequence is a set of things belonging next to each other on some principle of order, or alternatively a series without gaps. Both these definitions apply when it comes to the sequential launching of projects and for the following reasons:

- **Set of things:** The very principle of a programme is that it comprises a group of related projects and activities that together achieve outcomes and realise benefits that are of strategic importance;

- **Next to each other:** Sub-projects in a programme do not necessarily have to be directly adjacent to one another but they are linked and contribute to meeting the business objectives. Sub-projects can cover different facilities at different geographic sites;

- **Principle of order:** This is an essential component of obtaining maximum benefit from a programme. Sub-projects have to start-up in a logical order to ensure that the necessary utilities and infrastructure are available as required by other projects, and;

- **Series without gaps:** To keep the programme schedule to a minimum, sub-projects must be commissioned either concurrently or with no time gaps in between.

Principle of order

A power station cannot be commissioned without a ready supply of water and energy, i.e. coal or gas. A gasification plant cannot be commissioned without a supply of oxygen and steam. A refinery product work-up area cannot be commissioned without feed material and a supply of hydrogen.

There will always be one best sequence for projects to be commissioned in for any programme which contains a number of sub-projects. The principle of order is that facilities and plants should be commissioned in such a sequence that its deliverables are available in time to provide input or feed to the next one to be started. Some sub-projects can be commissioned concurrently.

This is probably all starting to sound familiar. Planning for the start-up sequence can be done using Gantt charts, such as Figure 7.6. A Gantt chart is a type of bar chart, developed by Henry Gantt in the 1910s, that illustrates a project schedule. Gantt charts illustrate the start and finish dates of the terminal elements and summary elements of a project. Terminal elements and summary elements comprise the work breakdown structure of the project.

By scheduling the sub-projects on a Gantt chart, it can be determined which project should start-up when in order to meet the overall business objectives.

Concluding Remarks

In this chapter we've discussed benefit management and the sequential launching of projects.

Keeping track of the objectives of the programme and measuring the improvements as the sub-projects become live will ensure focus on the end goals during the life of the programme, provide feedback and ensure engagement by all stakeholders. It also enables easy close-out of the programme without having to try and measure the benefits retrospectively and at great cost. It allows for close-out without accompanying emotions, blame and accusations amongst the different parties in the event that the programme has not delivered on the business objectives.

In the next chapter we will discuss the need for regular review of a programme.

Chapter 9:
Review Programme Performance

"I think it is an immutable law in business that words are words, explanations are explanations, promises are promises - but only performance is reality." - Harold S. Geneen

Introduction

A review is a structured opportunity for reflection to identify key issues and concerns relating to a project or programme at agreed intervals. It allows the team to make informed decisions for effective project/programme implementation. Review is the sixth element of the six-bladed propeller in our programme management model as indicated in Figure 9.1.

Programme review is a process that is normally initiated when managers and other stakeholders pause to assess how a programme has performed during a given period of time. A programme review is an integral part of the programme cycle. It is a form of programme monitoring that aims to provide feedback on performance of a programme to inform planning and improve implementation. Programme reviews build on routine programme monitoring and control. Three types of reviews are generally done on programmes. These are annual reviews, mid-term reviews and end-term reviews.

Throughout this book and especially the chapters on programme control and sequential launch (Chapters 6 & 7) we have been advocating continual transparency on where the programme is heading in terms of cost, schedule, quality and meeting the programme objectives. Regular tracking and reporting progress against these parameters as the programme continues, provides transparency to all stakeholders and enables corrective action to be taken as soon as a deviation becomes apparent. Setting up the necessary tools to be able to do so at the beginning of the programme enables this type of regular reporting to happen without much additional effort. When the leaders involved understand and buy into the concept, it becomes part of the day-to-day running of the programme and the sub-projects. Every programme update then enables a review of progress to date with stakeholders.

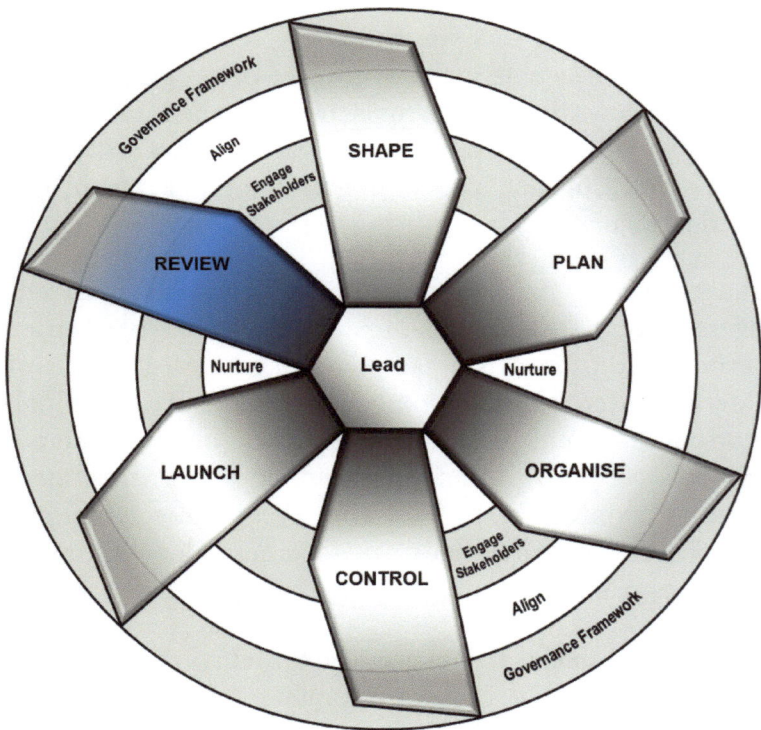

Figure 9.1: Programme management model with 'Review' highlighted

⚠️

Regular tracking of cost, schedule, quality and the extent that programme objectives are met, and reporting progress against these parameters as the programme continues, provides transparency to all stakeholders and enables corrective action to be taken as soon as a deviation becomes apparent.

Monitoring, reviews, evaluations and audits

Introduction

The words monitoring, review, evaluation and audit are sometimes used as synonyms, although they have different meanings. A comparison of the key features of each in the context of a project or programme is given in Table 9.1 below.

Table 9.1: Key features of monitoring, reviews, evaluations and audits

Feature	Monitoring	Reviews	Evaluations	Audits
Reason	Monitor progress, inform decisions and remedial action, support accountability	Check progress, inform decisions and remedial action, update project plans, support accountability	Assess progress and worth, identify lessons and recommendations for longer-term planning and organisational learning; provide accountability	Ensure compliance and provide assurance and accountability

Feature	Monitoring	Reviews	Evaluations	Audits
Timing	Ongoing during the project/programme	Regularly scheduled intervals during the project/ programme	Periodic, but typically after the project or programme has been completed	According to owner and/or financier requirements
Involved	Internal, involving project and programme implementers	Internal, involving project and programme implementers	Can be internal or external to the organisation	Typically external to project or programme, but internal or external to organisation
Focus	Focus on inputs, activities, outputs and shorter-term outcomes	Focus on inputs, activities, outputs and shorter-term outcomes and business objectives	Focus on outcomes and overall business objectives	Focus on inputs, activities and outputs against a standard

We use 'monitor' to reflect the continual gathering of relevant data regarding the programme in terms of expenditure, safety performance, percentage completion, performance against schedule and scope changes. The data gathered through the monitoring process is refined and packaged for the regular overall programme performance reviews.

Evaluations or review interventions

While monitoring is ongoing, reviews are less frequent, but not as involved as evaluations. Evaluations involve identifying and reflecting upon the effects of what has been done, and judging their worth. Their findings allow project and programme managers, beneficiaries, partners, financiers and other stakeholders to learn from the experience and improve future interventions.

In contrast to monthly progress meetings, an evaluation or review intervention is a major undertaking. The preparation for an evaluation can take up the full attention of the leadership team for a number of weeks. The resulting recommendations must be followed up afterwards and, worst of all, corrective actions can be futile as they are often after the event. Evaluation or review interventions are also often instigated because management becomes uncomfortable with the progress, mainly because of a lack of information. Programme and project managers should try to provide timeous information to eliminate the need for interventions and if company policy requires such interventions, then ensure that the normal programme practices are such that the required information will be readily available. This is indeed possible and can save much frustration and scurrying about.

Programme and project managers should try to provide timeous information to eliminate the need for evaluations or review interventions. If company policy requires such interventions, ensure that the normal programme practices are such that the required information will be readily available.

Project and programme audits

An audit is an assessment to verify compliance with established rules, regulations, procedures, standards or mandates. Audits can be distinguished from an evaluation in that emphasis is on assurance and compliance with requirements, rather than a judgement of worth. Financial audits provide assurance on financial records and practices, whereas performance audits focus on the three E's –

efficiency, economy and effectiveness of project or programme activities. Audits are typically performed by teams external to the project or programme, but can be internal or external to the owner organisation.

Seeing that project or programme audits are external to the programme, it will not be discussed further.

Importance of programme monitoring, reviews and evaluations

A well-functioning monitoring, review and evaluation system is a critical part of good project and programme management and accountability. Timely and reliable monitoring review and evaluation provides information to:

- **Support project/programme implementation** with accurate, evidence based reporting that informs management and decision-making;

- **Contribute to organisational learning and knowledge sharing** by reflecting upon and sharing experiences and lessons so that we can gain the full benefit from what we do and how we do it;

- **Uphold accountability and compliance** by demonstrating whether or not our work has been carried out as agreed and in compliance with established standards and with any other stakeholder requirements;

- **Provide opportunities for stakeholder feedback** to provide input into and perceptions of the work performed, and to allow the team to adapt to changing needs, and;

- **Promote and celebrate project/programme successes** by highlighting the accomplishments and achievements, building morale and contributing to resource mobilisation.

Monitoring and review reports

Keep reports brief

Stakeholders, especially executives, do not have the time or inclination to read a typical project progress report consisting of on average 50 pages. The need is for concise up-to-date information at a glance.

Campbell and Campbell (2013) in their book *The New One-Page Project Manager* argue that the status of any project, no matter how big or small, can be communicated on one single page. They make extensive use of visuals, graphics and colours to convey the relevant information in an easily understandable format to the stakeholders of the programme (for examples, see their website at. www.oppmi.com). The objective of their book is to explain how to communicate and manage any project with a single sheet of paper. They state that a key competency of any project manager is communication and the One-Page Project Manager (OPPM™) is a set of tools to enable him to effectively communicate upwards, downwards, and sideways. The OPPM™ is the central focal point at all progress review meetings and contains common terminology that enables a common language and thus better understanding.

The OPPM™ format covers the following on this single page:

- A very brief summary;

- The project objective and sub objectives;

- The main tasks to be performed that link to the objectives;

- The schedule associated with these tasks;

- Scheduled and completed tasks (including being on time, late or early);

- The persons accountable for each task;

- Risks and other qualitative metrics, and;

- Cost metrics.

We follow a somewhat similar approach in this book in that each key metric is illustrated graphically in a single view showing whether the element is on track or not. Every key metric includes a forecast of the end-state, whether it is end-of-job costs, estimated completion date or expected performance. If any deviations are foreseen, it also states what corrective actions will be taken. The information is not summarised on a single page, but that can be done following the guidelines of the OPPM™.

We return to the Intego case study to show the typical reporting format for the different metrics of specific interest in this programme. The following is an overview monthly report for the Intego programme used for overall communication.

Intego Environmental Compliance Programme (IECP)

Programme Review Report: November 2012

Programme Objective

Reduction of SOx, NOx and Particulate matter (PM-10) emissions by 10% in 2015, 20% in 2020 and 30% in 2025 as compared to the 2010 baseline.

Indicated Total cost

The indicated total (end-of-job) cost for IECP is given in Figure 9.2.

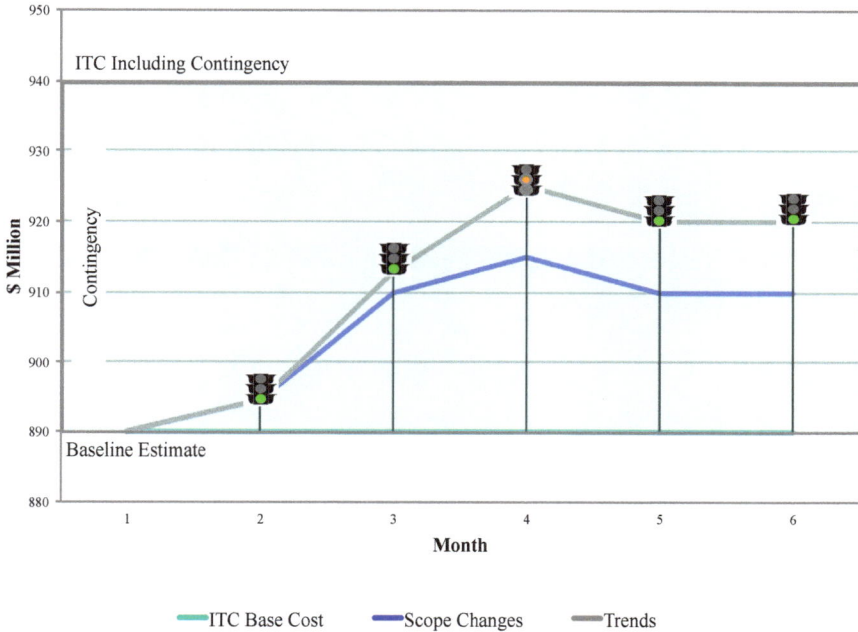

Figure 9.2: Indicated end-of-job cost for IECP

The additional nitrogen production unit contributed US$ 8 million with a payback period of less than 2 years and has been approved by the IECP Steering Committee for inclusion.

An increase of US$ 12 million has been caused by additional equipment for water clarification at Coal 1 as it was found that the water quality is actually worse than originally assumed. A value engineering exercise will be conducted to optimise the water purification system design.

Steel price increases has resulted in a trend forecast increase of US$ 10 million. This is due to a greater global demand for steel.

Further steel price increases are expected and the procurement department has been requested to investigate fixing steel prices by placing early orders on steel mills or forward buying of bulk steel.

Master Schedule

The master schedule for IECP is shown in Figure 9.3, showing which projects of milestones are ahead of, or behind, schedule.

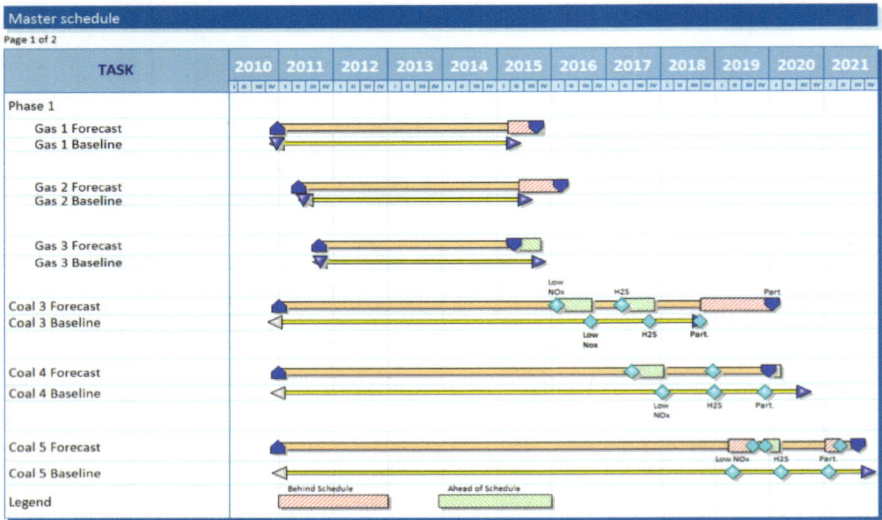

Figure 9.3: Master schedule for IECP

On Gas 1 and Gas 2 the detailed schedules will be reviewed to establish the actual cause of the predicted delay and activities will be adjusted to reduce the delay.

The cause of poor productivity on Coal 3 will be subjected to a root cause analysis and steps will be implemented to address the causes identified.

Programme performance

The historic and expected future performance of the IECP is

reflected in Figure 9.4.

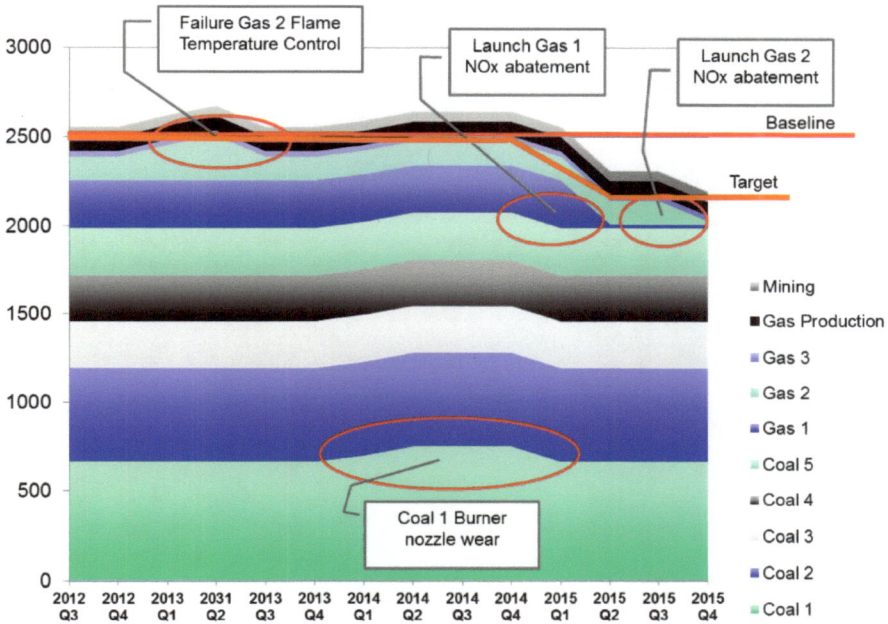

Figure 9.4: NOx emissions report

For illustrative purposes, the programme's objective of end 2015 for NOx only is used as a representative example (see the project charter example in Chapter 14 on governance for the full set of objectives).

The reduction in NOx emissions from the launching of the abatement projects on Gas 1 and 2 have delivered the desired results.

Risk management

Primary risk elements for IECP are included in Table 9.2. The table uses green to signify that everything is on track, orange to indicate that concerns have been identified and red to indicate that a problem has been encountered.

Table 9.2: Status of primary risk elements for IECP

Risk Element	Month 1	Month 2	Month 3	Month 4	Month 5	Month 6
Meeting shutdown schedules						
Labour stability						
Quality of engineering design						
Quality of construction	Not yet Applicable	Not yet Applicable	Not yet Applicable	Not yet Applicable	Not yet Applicable	Not yet Applicable
Brownfield risk						
Safety						

The problem that arose with the quality of engineering design during month 5 has been addressed. The quality of engineering design work has improved with the sourcing of experienced supervisors. At least another month will be required to stabilise the situation.

The accuracy of as-built drawings has been found to be wanting. Laser scanning techniques are being used to accurately document the as-built status. The exercise will take 6 months to complete.

Although the programme review report as presented here covers three and a half pages, there is certainly opportunity to reduce the size of diagrams and use bulleted notes to get closer to the ideal of a one-page report.

Concluding remarks

Depending on the company requirements, certain end-of-term or end-of-project-phase evaluations may be necessary and should be done to formally conclude and document the status at that point. If the guidelines in this book are followed, all the necessary information should be available for the programme and project documentation. As a programme usually consists of many interdependent projects, each project can be individually reviewed at the end of each phase or after all projects in the full programme have been in operation successfully for a period (usually about 6 months).

An end-of-programme evaluation is recommended to capture lessons learnt from the successes and failures of the programme. This will help to prevent the same errors in future and help make the owner organisation a learning organisation.

This brings us to the end of part 2 of the book where we discussed the six sequential steps of programme management: the six-bladed propeller part of the programme management model. In the next part, we tackle the programme lifecycle essentials or those issues that need to be in place if you want your programme to be successful. Programme lifecycle essentials are indicated in the target part of the programme management model.

This page intentionally left blank

PART 3
Lifecycle Essentials

This page intentionally left blank

Chapter 10:
Leadership for Success

"Leadership is solving problems. The day soldiers stop bringing you their problems, is the day you have stopped leading them. They have either lost confidence that you can help or concluded you do not care. Either case is a failure of leadership". - Colin Powell

Introduction

Leadership is one of the most important lifecycle essentials for project and programme success and therefore it occupies the central position in our programme management model as shown in Figure 10.1.

Project managers are all in leadership positions, but not all project managers are leaders. Project leadership is very different from project management. A project manager can be described as the person responsible for directing and coordinating human and material resources, but this definition tends to focus on the administrative aspects of project work. Verma and Wideman (2002) see a distinction between the style of leaders and managers according to their primary focus, as listed in Table 10.1. They consider it a truism that leaders focus on doing 'the right things' while managers focus on doing "things right". The sponsor of a programme is a leader guiding the team to achieve business success, whilst the project managers lead their teams to project success.

In the context of a programme, the leadership traits are more desirable for the top positions, namely the sponsor and the programme director. Managers of the sub-projects will do well if they demonstrate the managerial traits, but will do better if they also demonstrate leadership qualities. Jordan (2009) says that one should not confuse leadership with authority: they aren't the same. He expects an effective leader to be able to inspire and motivate their teams, to develop their resources to be more effective contributors, to understand how their work fits in to the bigger picture and to be able to make the tough decisions when necessary.

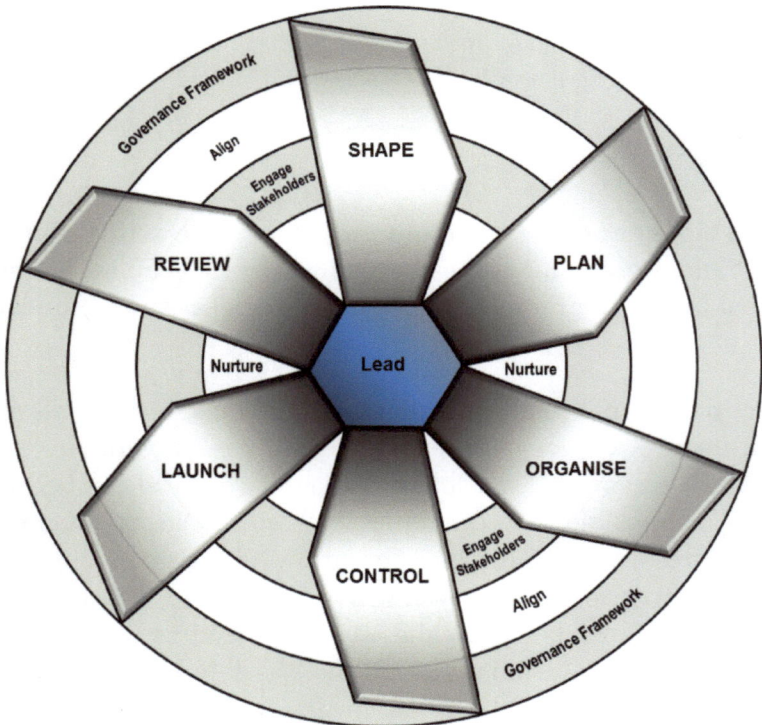

Figure 10.1: Programme management model with the 'Lead' element highlighted

In this chapter we briefly review leadership and project leadership, debate the concept of project stewardship, look at different ways to improve leadership on projects and discuss how to effectively implement change.

Table 10.1: Distinction between managers and leaders

Managers focus on:	Leaders focus on:
• Goals and objectives	• Vision
• Telling how and when	• Selling what and why
• Shorter range	• Longer range
• Organisation & schedule	• People
• Autocracy	• Democracy
• Restraining	• Enabling
• Maintaining	• Developing
• Conforming	• Challenging
• Imitating	• Originating
• Administrating	• Innovating
• Directing & controlling	• Inspiring trust
• Procedures	• Policy
• Consistency	• Flexibility
• Risk avoidance	• Risk opportunity
• Bottom line	• Top line

Leadership and project leadership

Defining leadership

Before considering what leadership is, it makes sense to say what it is not:

• Leadership has nothing to do with seniority or one's position in the hierarchy of a company;

• Leadership has nothing to do with titles;

- Leadership has nothing to do with personal attributes, and;

- Leadership isn't management; managers manage things whereas leaders lead people.

Kruse (2013) defines leadership as a process of social influence, which maximises the efforts of others, towards the achievement of a goal.

Key elements of this definition include:

- Leadership stems from social influence, not authority or power;

- Leadership requires others, and that implies they don't need to be direct reports;

- No mention of personality traits, attributes, or even a title; there are many styles and many paths, to effective leadership;

- It includes a goal, not influence with no intended outcome, and;

- It includes the 'maximisation of effort'.

Now we turn to project and programme leadership.

Defining project and programme leadership

It has become accepted over the last 40 years that a range of leadership skills are required for effective project and programme leadership. Effective programme leadership, in turn, significantly increases the probability of a successful programme. The leadership skills for effective programme leadership include:

- Visioning;

- Influencing;

- Communicating;

- Listening and questioning;

- Strategising;

- Empowering, and;

- Team building.

Most of the skills listed highlight the importance of 'people' in the project and programme environment.

With these skills in mind, Verma and Wideman (2002) define project leadership as the ability to get things done well through others. According to them, project leadership requires:

- A vision of the destination;

- A compelling reason to get there;

- A realistic timetable, and;

- A capacity to attract a willing team.

The same definition applies to programme leadership. For programmes, because of the extended schedules, complexity and uncertainty, leadership is of even greater importance. It can be seen from this definition that the role of creating a vision and a compelling reason to get there belongs with the programme sponsor, whilst creating a realistic timetable and the capacity to attract a willing team resides with the programme director.

Project and programme leadership is the ability to get things done well through others.

Project stewardship

Verma and Wideman (2002) refer to project leadership and project 'managership' collectively as project 'stewardship'. To be a steward is to hold something in trust for another or to act as custodian. They define project stewardship as a willingness to be accountable for the well-being of the project organisation while placing service towards the goals of the project above self-interest. Stewardship entails holding accountability for your people without exacting harsh compliance from them.

In the planning phases of a project, 'managership', has limitations. These limitations are overcome by effective leadership. During the more structured construction or execution phases, leadership has its limitations, and here 'managership', or the project management role, is more appropriate. Companies therefore often use project managers with more prominent leadership capabilities during these early phases of a project and then hand the projects over to execution project managers at final approval.

Programmes, however, present a further challenge. While the initial conceptualising of the overall programme occurs at the beginning, the programme development and especially the planning of individual projects happens in parallel to the final execution of other projects, as pointed out before. This requires the programme leaders to be able to constantly focus on both the leadership and managership aspects of stewardship. However, as the work progresses, it is easy to start focusing on the management aspects only (as discussed in Chapters 4 to 9).

A graphical presentation of this concept is presented in Figure 10.2. For projects, the level of leadership required drops off shortly after the start of the project while the level of management required picks up to peak at the end. For programmes, the level of leadership required drops of much slower at first. Note that Figure 10.2 is purely illustrative and that for both projects and programmes the levels of leadership and management vary from start to finish, but remain relatively high.

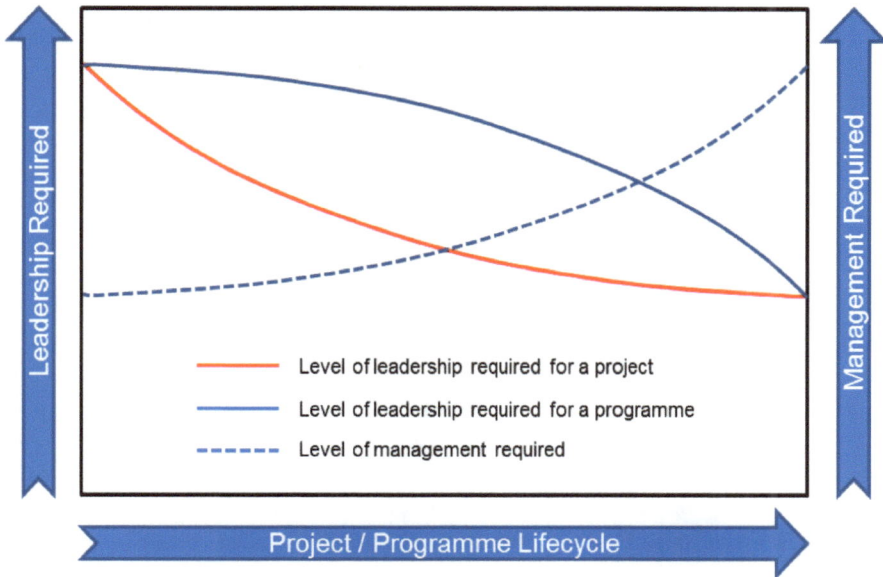

Figure 10.2: Requirement for leadership on projects and programmes

Programme managers and executive sponsors should therefore be aware that they are uniquely placed to make a huge positive or negative impact. This is where leadership skills and behaviours become critical: The programme manager must continually develop effective leadership and behavioural skills and employ them as needed in an environment without the benefit of formal line management authority.   Superior project leadership requires technical depth and management breadth. Programme leaders must possess a custom blend of specialised knowledge and key leadership skills to drive optimum programme performance. They need the ability to harness their team's collective intelligence, exercise appropriate influence, no matter what their hierarchical level, and communicate effectively.

The programme sponsor, especially, can play a major role in ensuring that the business objectives and progress towards it is constantly reinforced and communicated to executive management and company boards, the programme leadership team, as well as all other stakeholders. Reviewing the programme charter at regular intervals can be an effective tool in this regard. An effective communication plan (see chapter 12) is also an essential leadership tool. Effective communication will certainly affect the attitude of stakeholders towards the programme.

Ensuring proper project leadership

According to Mathis (2013), there are six ways to give proper leadership as you are setting up a project team, namely:

- **Create an atmosphere of trust:** Mutual trust amongst team members is an important feature of successful project teams. Trust can only be earned and starts off by consistently treating people in a respectful manner. Lies and backbiting kill trust and cause people to reject leadership. People can handle mistakes or even failure, but they cannot handle lies and disrespect;

- **Build the right team:** Most project teams experience some measure of interpersonal conflict during the project or programme lifecycle. Communication and common courtesy can break down causing the project to suffer. By carefully selecting the team members from the outset, some of these problems can be avoided. Once the right team is in place, implement a wellness management programme;

- **Spell everything out for your team upfront:** Start the project by telling all team members exactly what will be required from them. It always works better to tell people the truth. Explaining the complexity and scope of the project and how much time you anticipate it will take for completion will build your credibility. By giving the team the information up front, you set a tone of respect and courtesy;

- **Monitor and give feedback:** Giving proper feedback on the positives and negatives of a project is very important. If you think people are doing a great job, tell them. Positive reinforcement helps keep people focused on the right track. You also need to be willing to discuss where team members are lagging behind on the project. There must be a willingness to talk freely about whatever is needed to drive the project;

- **Keep communication open:** The creation of communication plans can assist in this area (see chapter 12). Avoid one-way communication which is only from top management downward. Communication is needed which crosses department lines, involves all stakeholders and keeps everyone informed and on board, and;

- **Keep the end goal in mind:** Project team members may lose focus during a project and allow deadlines to drift. Missed deadlines early in the project can have a major ripple effect. If not corrected, this ripple may continue until the end of the project. In the case of programmes where we plan for the sequential launching of project outcomes, this creates huge pressure for the programme team members at the later stages of the programme.

Building an effective team

Introduction

We have seen that one of the ways to ensure proper project or programme leadership is to build the right team. The programme director needs to ensure that he builds an effective and efficient team that can work together under difficult circumstances and for an extended period of time.

Identifying the competencies required and sourcing the individuals is a first step. On a large team as required for a programme, the individuals are typically not only from the owner organisation, but also from contractor organisations. Programme team members can thus have very different perspectives, cultures and backgrounds. A newly assembled team enters a forming phase where the individual members become orientated, need direction and seek acceptance within the team. It takes a while for the members to get to know each other, the requirements of the programme and the working methodology before they truly become a team.

Leading the team to high performance

It remains the task of the team leader to develop the new team into a well-performing team where individuals work productively towards shared goals, where open communication and trust is prevalent, where problems are solved as they crop up and conflict is handled in a mature way. Unfortunately, it is not easy for a team to reach the high performance level. To reach a high performance level requires a lot of preparatory work. Team development typically moves through four stages, as follows:

- **Forming:** Forming refers to the initial stage in the lifecycle of the team where vacancies in the programme team are filled and the working methodologies for the team are developed;

- **Storming:** Storming follows immediately after the forming stage and is characterised by confrontation, conflict and low productivity. Too many project teams get stuck in the storming phase. These teams can only deliver mediocre results;

- **Norming:** Norming describes the stage where the team starts to work together well and begins to produce results with relatively few interpersonal problems, and;

- **Performing:** This refers to the desired end state for the programme team, namely a high performance team where everyone knows exactly what is required and work towards the business objectives in a harmonious manner.

This progression through the stages does not happen naturally and needs constant attention from the leadership team. Leaders should never be satisfied with mediocrity and need to keep on pushing the team towards improved performance. It is a fact that a well-developed team provides superior results.

The four stages of team development is shown in Figure 10.3, together with the corresponding approach of the programme leader. The team's output increases drastically as it develops to the performing stage. Similarly, as the team becomes more effective, the input required from the team leader becomes less

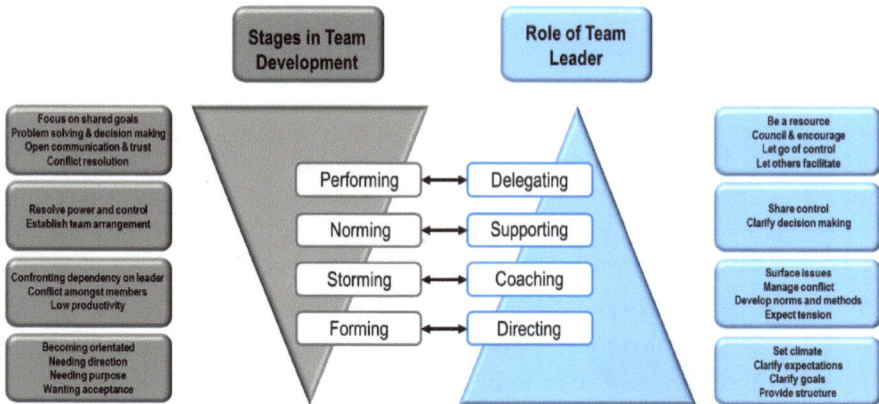

Figure 10.3: Team development stage and leader input

In order to build a well performing team it is firstly critical to ensure that the team has a common purpose, agreed performance goals (specific, measurable, assignable, realistic and time-related) and subscribes to a common approach supported by:

- Selecting team members based on skill and skill potential and not only on personality;

- Setting clear rules of behaviour;

- Paying particular attention to the first meetings and actions to set the basis;

- Clarifying expectations;

- Establishing urgency, and demanding performance;

- Setting a few immediate performance tasks and goals;

- Spending time together, and;

- Exploiting the power of positive feedback, recognition and reward.

Implementing change effectively

Introduction

A programme is by its very definition a major intervention for the owner company. Implementing the change implies that the programme leadership team also needs to be effective change agents. Resistance to change does not normally come from the project team members, but rather those in the organisation that will be impacted. Unless a positive and supportive environment can be created, the chance of programme success will be greatly curtailed.

The following are key elements of change management that the sponsor and programme director need to enforce:

- Continually reminding the team of the vision and the reasons for change;

- Rewarding and publicising milestones achieved;
- Effective communication, to:
  - Share the vision, progress, recognition;
  - Get feedback on progress and areas of concern;
  - Show understanding, and;
  - Report progress to the programme board.
- Understand the political power play, the backers, opponents, supporters and detractors and the approach towards them.

Programmes are long-term initiatives

All too often, focus is placed on the programme and project teams, but the actual personnel in the owner organisation where the change will be implemented are largely ignored. Suddenly they are faced with this new initiative and the typical response is denial, fear, resistance and potentially conflict.

In contrast to a single project, leading a programme is a longer-term initiative and the focus should be on the following five levers of programme leadership:

- **Strategy:** The overall programme strategy should be clear and well communicated. The strategy is generally laid out in the programme charter

  Strategy should be a central, integrated, concept of how the programme will achieve its objectives and includes:
  - Where we want to be?
  - How will we get there?
  - How will we define success?, and;

o   What is the speed and sequence of moves?

- **Structure:** Structure refers to the roles and relationships that exist in the organisation structure for the programme team. The structure of the owner organisation itself can also play a role. The following issues deserve attention:

  o   Basis of the organisational structures, systems and tools;

  o   An emphasis on stimulating lateral or innovative thinking and what is required to achieve this;

  o   Deciding whether the programme team and thus the various sub-project teams will be co-located or in disperse locations, and;

  o   Deciding how centralisation of procedures and decisions at programme team level will be handled vs decentralisation at sub-project team level.

- **Information and decision processes:** This refers to the flow that occurs within the programme team and externally to the stakeholders. The more pertinent processes include:

  o   Defining the decision making approach in terms of levels of authority and approval process;

  o   Defining planning, goal setting and monitoring requirements and procedures;

  o   Setting up the necessary controls and metrics for feedback on progress;

  o   Developing communication requirements in terms of content, format, distribution, and;

  o   Agreeing responsibilities for continual and pro-active scanning of the external environment to ensure early knowledge of all aspects that could impose new requirements or restrictions on the programme.

- **Rewards:** Rewards cover those mechanisms for stimulating and propelling human behaviour (i.e. getting the right people to join, to stay, and to perform at their fullest), and includes:

    o Defining the types of rewards that will be used, be it financial or intrinsic recognition programmes, taking into account what happens in the rest of the organisation;

    o Defining criteria for receiving rewards: Will it be based on individual, team, affected business unit, whole programme or even whole company performance, or a mixture;

    o Establishing a balance between long-term vs. short-term measures, and;

    o Determining how the development, promotions and advancement of individual team members will be handled and by whom.

- **People:** People refers to the composition of the team and the mix of human talent that exists in the programme team such that all the required competencies are included and covers issues such as:

    o Procedures and guidelines for recruitment and selection;

    o Defining job content and selecting individuals with the ability to apply their perceived competencies;

    o Defining processes for performance evaluation, coaching and leadership development;

    o Agreeing how job rotation (cross function, outside and within the programme) can be used to develop and motivate personnel, and;

o   Ensuring training and development is taken care of, including types of training and whether hard and/or soft skills are to be included.

Concluding Remarks

The importance of effective leadership for programmes cannot be overemphasised and is a prerequisite for programme success. The most important role that a programme leader has is to grow a recently formed programme management team into a high performance team.

It is beyond the scope of this book to go into the detail of effective leadership. We recommend that readers do their own research into this fascinating topic. Two interesting books with a somewhat different slant on the topic are:

*   *Provocative Therapy, 5th ed.* by Farrelly and Brandsma (1989): In this book the authors emphasise the potential benefit of being deliberately provocative rather than being accommodating. The situation will dictate which approach is the most appropriate, and;

*   *When Sparks Fly: Harnessing the power of group creativity* by Barton and Swap (2005): The authors describe a method that can help people become more innovative and better at teamwork. The process involves five steps: selecting the right mix of people to spark creativity; identifying the problem needing novel ideas; developing alternatives; taking time to consider choices; and selecting one option.

In the next chapter we consider the basic human needs of the programme team and discuss the concept of nurturing.

Chapter 11:
Nurturing the Programme Team

"Really in technology, it's about the people, getting the best people, retaining them, nurturing a creative environment and helping to find a way to innovate." - Marissa Mayer

Introduction

According to the Oxford Dictionary, nurturing means to "care for and protect (someone or something) while they are growing". In a programme context, nurturing refers to the care and protection of team members during the extended duration of a programme. Nurturing is the second supporting process in the programme lifecycle essentials part of the programme management model as shown in Figure 11.1.

Nurturing remains important throughout the lifecycle of a programme, as is the case with the other programme lifecycle essentials (leadership, stakeholder engagement, alignment and governance). In reality, a programme provides for a high stress environment, leading to fatigue. Fatigue again leads to poor performance and an increased potential for incidents. Programme managers should be acutely aware of this and plan to manage the consequences pro-actively.

On a programme, nurturing covers both the working and personal environments of team members. Nurturing considers aspects like personal development, recognition, wellness, both on a team and individual level, as well as personal health and fitness. Nurturing also takes into account an individual's personality, tolerance to stress and reaction to stress. These aspects and the interrelationship thereof are shown in Figure 11.2. In order to build the effectiveness of the integrated team, each and every team member needs to be nurtured.

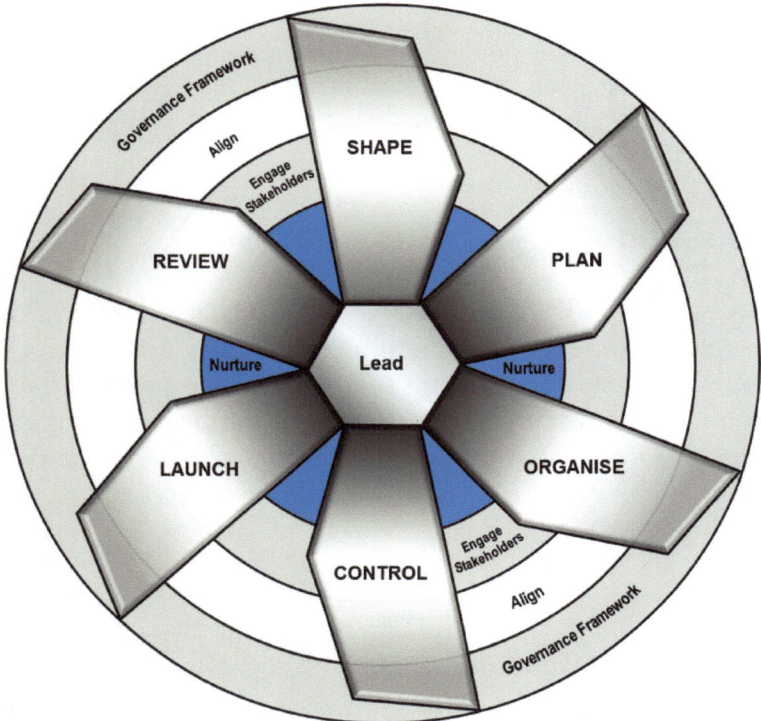

Figure 11.1:   Programme management model with the 'Nurture' element highlighted

A significant part of nurturing on a programme focuses on team wellness; alternatively it can be called nurturing a team. To do this, one needs to see team wellness in context. Simplistically stated, team wellness is dependant of two major factors, namely the work environment of the team and the personal environment of the individuals on the programme team. The environment in which the person lives outside work, and operates in whilst at work, has a direct impact on team wellness. Are these environments conducive to a high performing team? Does it bring about positive stress (eustress) or negative stress (distress)?

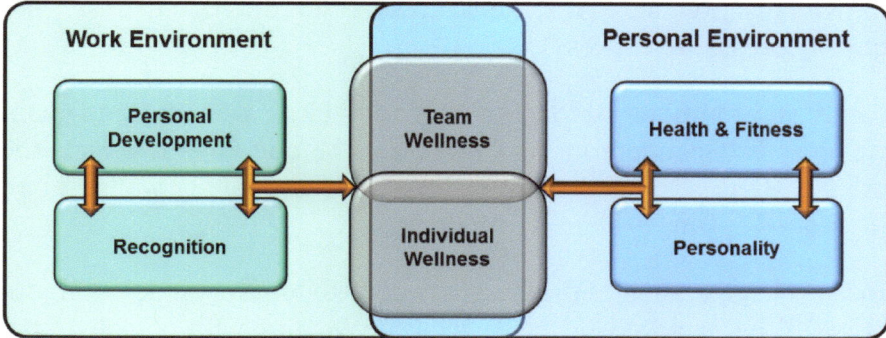

Figure 11.2: Nurturing in context

According to Mills, Reiss and Dombeck (undated), stress is not always a bad thing. They continue that eustress, or positive stress, is optimal stress that motivates a person, is short-term in nature, is stress within your coping capability, helps you feel excited and heightens your levels of alertness. All this contributes to improved performance. In contrast, distress, or negative stress causes anxiety or concern, can lead to mental and physical problems if greater than what we can cope with, and ultimately decreases performance. Distress is also more long-term in nature. We argue that there is a causal effect between the long-term pressures of a programme and distress.

Stressors emanating from your work and your personal environments affect individual and team wellness. Key elements contributing either to eustress or distress in the work environment is the extent to which a team member feels valued and recognised, and is given the opportunity to grow and develop. Further contributing to eustress or distress in the personal environment is the extent to which the individuals look after their health through eating well, exercising and sleeping enough. How a person copes with stress is greatly influenced by his/her personality. Certain personality types tend to put more pressure on themselves and others, and if the individual's tolerance level for stress is exceeded, eustress very quickly turns into distress.

Lastly it is important to note that individual wellness and team wellness is dependent on one another. The one directly affects the other. Hence, in addressing distress, and ultimately wellness, an integrated approach should be followed.

In this chapter, each of the elements depicted in the model in Figure 11.2 will be explored in more detail. We start with a look at the personal environment, then the work environment and lastly focus on an integrated wellness programme.

Individual Wellness

Introduction

Even though there is no clear average lifetime of a programme, programmes can typically last between 5 to 20 years. At the onset of a programme, an integrated wellness programme should be put in place. During the very early phases of a programme, it may be premature to invest money into a wellness programme but the planning should be done to be ready to implement the wellness programme as soon as the programme charter is agreed. Provision should also be made in the programme budget for a wellness support programme. We'll go into far more detail later in this chapter on integrated wellness programmes.

According to Imtiaz and Ahmad (2009), stress is a universal experience in the life of each and every employee, including executives and managers. It is estimated that about 100 million workdays are being lost due to distress and 50% to 75% of diseases are related to distress (Bashir, 2007). Distress results in a high portion of absence and loss of employment. The percentage of stress-affected employees in organisations is increasing at an alarming rate, affecting both employee performance and corporate goal achievement (Treven, 2002).

Impact of programmes on individuals

Anyone that has ever worked on a megaproject would know that the constant pressure to achieve cost, schedule and quality targets can be daunting at times. Most individuals handle pressure and job stress remarkably well, but what makes it worse on a programme is the prolonged nature thereof.

Programmes tend to impact individual employees in the following manner:

- Their careers may be seen as stagnant, since they stay in a certain position for a long period of time. Hence employees tend to become demotivated and disengaged;

- Programmes typically calls for a lot of travelling. Inevitably fatigue as a result of extensive driving and flying sets in;

- Temporary offices and site conditions may be less than ideal and overcrowded and unpleasant working conditions further contribute to work stress;

- Long periods away from home and the family may cause relationship problems and the breakup of marriages, and;

- When traveling and working long hours, employees tend to neglect their health and fitness.

The build-up of work stress becomes a vicious cycle. One variable starts impacting on the others, and very soon performance levels drop and matters get out of control. The best way to approach nurturing on a programme is to introduce an integrated wellness programme prior to the symptoms of poor performance, illness and disengagement becoming evident. Integrated wellness programmes are discussed in more detail in the section on team wellness.

Personality

According to Forrester (2012), your personality affects the way you cope with stressors in your life. Some people cope fine with levels of stress that others would find crippling. Our reactions to stress are partly inborn and partly learned throughout our lifetime. Everyone reacts to stress; however, it depends on the level of stress you can handle before symptoms arise.

Stress builds up until it reaches a critical level and symptoms appear. Finding a suitable way to relieve stress will assist you to cope better. There are many different ways to reduce stress, but the way that will work best for you depends on your personality. We all need to consider our stress levels and work consistently to reduce them, because stress related illnesses are common and emotional problems caused by poorly controlled stress levels can affect us all, irrespective of whether we are the sufferer ourselves or affected by a colleague or relative's problem.

Forrester (2012) lists the following four personality types with the appropriate stress reduction technique:

- **Perfectionists** create their own stress by being too hard on themselves. They need to be more realistic and forgiving of themselves. Perfectionists should build leisure time and relaxation breaks into their daily routines;

- **Thrill-seekers** thrive on risk. Their attention span can be short as they are racing on to the next thing. These people tend to smoke or drink too much and forget to take care of themselves properly. Often these types respond well to taking part in energetic sports, to burn off the tension;

- **Anxious** types have poor self-esteem. They often take on jobs that are not demanding enough and then become bored and frustrated. Although relaxation and meditation is helpful for this group, it should be combined with methods to build self-esteem, and;

- **Ambitious** personalities are often the most stressed, but least ready to do anything about it! They deny that they are stressed and just keep going until ill health forces them to stop. These are power driven people, often somewhat aggressive when crossed. A combination of relaxation and active leisure pursuits can help here.

Although different personality types require different techniques to reduce stress, Creagan (2012) states that stress can change your personality, mostly not for the better. Stress can erode your spirit and decrease your quality of life, but you may not even be aware of it. Such a situation will obviously have an adverse effect on job performance.

Health and fitness for the job

This heading is deceptive on purpose. We want programme team members to be fit healthy and happy to be able to perform their tasks optimally. Health concerns, naturally, are a big drain on an employee's ability to be productive. As we have seen, stress is an important trigger for mental and physical health concerns.

The second implication of the heading is to appoint team members who are suited to what is going to be demanded from them. Two simple examples are:

- Do not to appoint team members who cannot interact well with others in programme leadership positions, and;

- Do not appoint artisans who suffer from vertigo when they are expected to work at heights.

In essence, this means that every job has to be evaluated in terms of what personality types and medical profile can be tolerated. Candidates will then have to be screened carefully to ensure that suitable appointments are made. This will prevent many incidents and future dissatisfaction.

Team Wellness

Introduction

A key success factor on any project or programme is consistency of human resources. Due to the long duration of programmes, consistency of resources is even more critical on a programme. To work on a programme for 10 years in the same position would not motivate anyone. There are some strategies one can employ to manage this dilemma. Personal development and recognition are critical levers that programme directors should use to motivate team members.

Personal development

A useful strategy for developing promising individuals is job rotation. The programme would require proper succession planning in a structured way that allows new team members to take over. It benefits the individual since they gain exposure over a wide range of functions and business areas.

Another way to keep team members content, is to ensure that they are not overseen for promotion. On programmes, key individuals tend to play a critical role over time seeing that they get to know the background of the programme, assumptions underlying the programme and contextual issues that may have impacted the programme. Management tends to keep these individuals in their particular roles for extended periods without considering the growth aspirations of the individual.

It is recommended to have a flatter career band approach on programmes, rather than a multiple level approach. A person may thus remain in their post band for multiple years, but can receive performance based increases within that band at any given time. To avoid individuals being overseen, a formal review of salaries should be done annually in conjunction with the performance appraisal cycle.

Retention bonuses can also be considered. Monetary reward is a great way of motivating team members, but non-monetary recognition and acknowledgement can also be a powerful motivation tool.

Recognition

Programmes should always make provision for recognition, both informal and formal. Because employees want to feel valued and appreciated, informal recognition of employees' accomplishments has the potential to positively impact both individual workers and the organisation as a whole. Informal recognition can be described as the use of positive reinforcement without material rewards. For example, informal recognition can consist of a simple note or a verbal 'thank you'. It differs from formal recognition, in which rewards such as gifts and money are granted. A way should be found to establish informal recognition as the preferred culture on a programme.

Formal recognition, on the other hand, should be done on a continuous basis with intervals varying between one and four months. Formal recognition systems should be linked to the programme drivers and team values you want to encourage. These can include safety, quality, cost consciousness, team work, environmental performance, excellence and so forth. It is also advisable that as the programme changes, and priorities or drivers change, that these focus areas be adapted to suite the direction of the programme. They need not be static over the lifespan of a programme. The mistake so often made is to run recognition as a top down approach. The greatest 'programme heroes' are often not visible to management, but rather seen by the workers on the shop floor. Therefore the principle should be that any team member should be able to nominate any other team member on any level in the organisation. Recognition is, however, never a quick fix. Results are typically only visible after six months of consistent application. What works best on projects is a combination of top-down recognition, peer-to-peer recognition and bottom-up recognition.

Effective and consistent management of a recognition system will be the ultimate determining factor if it will achieve its goal of motivating the project and programme teams. Once a recognition system is kicked off, it should be diligently administered. The sooner the recognition is given after the event, the greater the impact. Do not slip up for a few months due to other pressures. Appoint someone to review the nominations and provide a recommendation to the executive team. This administrator will also ensure that awards are handed out timeously and that it is published on the programme intranet site the moment it has been communicated.

As for everything else on a programme, the cycle of nomination submission, deciding on the best submissions for the month, communication and reward should work like a well-oiled clock.

Personal development and recognition for the Intego programme

Here follows a brief summary of the personal development and recognition interventions taken on the Intego Environmental Compliance Programme.

Intego Environmental Compliance Programme (IECP)

Personal development and recognition

Personal Development

On the Intego Programme, all young professionals with less than two years' working experience were rotated in order to provide them with exposure in key areas of the business, as well as different disciplines. They spend an average of 12 to 24 months in any given area. Less than 12 months is not enough time for them to really get to grips with the environment and to really make a difference. Transfer and promotion from one sub-project onto another sub-project with more responsibility within the programme was encouraged. When new resources joined the programme, a special effort was made to understand their development needs and career aspirations and, via exposure to the programme, accommodate their aspirations. To this effect, a full time people development manager was assigned to the programme.

Recognition

For the formal recognition system, emphasis was placed on:

- safety from the perspective that team members should look out for one another's safety and report unsafe acts as a pro-active measure to avoid incidents;
- creative, yet cost effective solutions to complex problems;
- on-time delivery without negatively impacting defined quality requirements, and;
- team work within teams, but also across teams and businesses within the Intego structure.

Due to the long-term duration of the programme, a peer recognition system was introduced. The recognition initiative was called: INSPIRE ME. All team members have access to an intranet-based template where nominations can be logged at any time and recorded onto a single database. See Figure 11.3 for the template in use for recording nominations for recognition. All nominations are posted on the intranet on the INSPIRE ME recognition page.

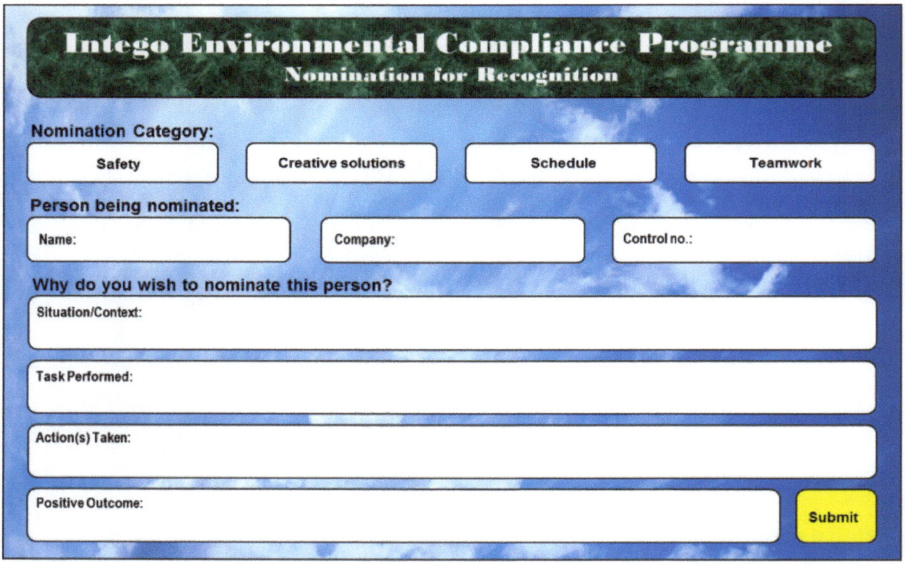
Intego Environmental Compliance Programme
Nomination for Recognition

Nomination Category:
Safety
Creative solutions
Schedule
Teamwork

Person being nominated:
Name:
Company:
Control no.:

Why do you wish to nominate this person?
Situation/Context:
Task Performed:
Action(s) Taken:
Positive Outcome:
Submit

Figure 11.3: Intego recognition template

Though submissions are done monthly, one nominee is chosen quarterly as the 'Star of the Quarter'. The quarterly winners are recognised with a dinner voucher for his or her immediate family. The actual cost of the dinner was covered. No limit is defined, only the requirement that it only covers direct family members. Once a year, the 'Star of the Year' is selected from all the nominees, and recognised with a weekend holiday voucher for the immediate family.

The system is easy to administer and has a significant impact on the morale of team members. It is, however, important to have an objective and fair team who reviews the submissions.

Retention bonus

Over and above the usual mechanisms discussed above, a retention bonus is used. When key programme team members joined, they were signed up for the retention bonus, which meant if they stayed with the programme for 3 years, they would receive a substantial bonus which equates to 8 month's salary. This was only done for pre-defined key roles on the programme.

Changed programme targets

Team targets are sometimes changed depending on the strategic focus required at any particular time. At one stage an incident occurred that could have resulted in a fatality. A team incentive was then launched for consistently demonstrating safe behaviour and for initiative by teams to get safety measures in place that visibly reduced safety risks.

At another time, financial pressure due to an economic downturn led to new challenges for the programme. An efficiency drive was launched where teams had the opportunity to submit ideas on how to improve the efficiency of any process within their sub-projects, which could result in cost or time savings without compromising quality specifications.

Such campaigns should not last less than 6 months. A good duration is 6-12 months. The ideas should first translate themselves into tangible results. The winning teams and their spouses were treated to a weekend away at a local resort. To avoid demotivating other teams, the programme management team has to ensure that the winning team is indeed the team that delivered the best results in terms of the defined criteria.

Integrated Wellness Programmes

Introduction

According to Treven (2002), wellness programmes have been developed to help employees to maintain their physical and mental health. These programmes are also known as employee assistance programmes (EAPs). A healthy person has a greater capacity to manage stress than the one suffering from phobias, nightmares, lack of appetite, heart disease or other health problems. Typically, wellness programmes consist of workshops that train employees on how to pro-actively manage their stress, such as losing weight, exercising, giving up smoking and the like. It also includes support in the form of therapy and counselling. Although organisations provide the relevant knowhow, the individual employees are responsible for taking control over their own lives. Organisations that provide wellness programmes for their employees consider such programmes a sensible investment with a positive financial impact. The employees who are able to manage distress effectively enjoy better health, which in turn means reduced absenteeism from work.

On a programme the approach to wellness programmes is slightly different. The first step in the wellness programme should be a baseline assessment of all programme team members. There are many stress or burnout questionnaires available, which can be used to determine the emotional and psychological health of employees, as well as their proneness to burnout and unsafe behaviour. It is essential to engage a professional company, which makes use of certified and experienced personnel to do the assessments and follow-up discussions. It is also critical that the service provider acts responsibly with personal information and that all information on individuals is kept confidential. Reporting to the employer is done using dashboards of the overall wellness indicators of the team in terms of numbers of cases. Reports reveal no personal details, unless the employee consents thereto.

Stress assessments

Stress or burnout surveys typically determine team members' level of engagement or disengagement. Much research has been done on employees and their levels of engagement and how that affects their performance. According to Wikipedia, an 'engaged employee' is one who is fully absorbed by and enthusiastic about his/her work and takes positive action to further the organisation's reputation and interests.

Under continuous high levels of pressure, having personal problems at home, not feeling physically well and not feeling appreciated, team members tend to become disengaged. Many factors, both in the workplace and in the home environment, and more often than not, a combination thereof, can contribute to disengagement. This leads to unsocial behaviour, outbursts, family problems, poor health, heart attacks, conflict in the work environment, substance abuse, and ultimately low productivity and unsafe behaviour.

In supporting team members using a wellness programme, a baseline assessment and re-assessment at least every six months is required, as an individuals' profile may change over time. Service providers for these assessments should be selected based on competency, their professionalism, and experience. The assessment tool and the process to follow up on the findings should make provision to follow up on both the work environment and the personal environment. Issues relating to the work environment should be reported to management. Sometimes a single personal stressor can cause immense pressure on an individual's capacity to perform effectively. Two of the most common examples are when an individual has a sick family member at home whom they need to look after, or if they have financial problems. In the first instance such an individual does not sleep well, is worried the whole day and battles to focus. In both instances, wellness programmes can support such an individual by, for example:

- Counselling and advice on how to deal with the issue;

- Advice on home care services;

- Introducing the individual to NGOs that provide home care, and;

- Financial life skills training.

Every new member joining the team should be encouraged to complete the baseline assessment as part of their formal orientation to the programme. It does, however, remain voluntary. A question to be answered is which employees should be included in the wellness programme? The easiest is to start at the top of the organisation with the management structures. But where does one stop? Do you involve all contractors as well? The simple answer is 'yes'. As part of the contracting process, all partners, contractors and key stakeholders who actively work on the programme should be commercially bound to participate in the wellness programme.

If the wellness programme is run well, it can prevent incidents and fatalities. This is especially true if applied to construction contractors. It is worth every cent to commit all stakeholders to participate in the wellness programme. The funding of participation in the wellness programme needs to be agreed at the outset and the requirements included in the contracts as they are being negotiated.

Intego Environmental Compliance Programme (IECP)

Integrated Wellness Management

Introduction

On the Intego programme it was decided to partner with a local university who has commercialised an employee engagement questionnaire. The questionnaire measures team members' levels of engagement and disengagement.

This approach proved to be relatively inexpensive, seeing that students were co-opted to do the assessments as part of their studies. The university also gave a special dispensation on the analysis of the assessments. A 48-hour turnaround time was agreed. However, if signs of unsafe behaviour or burnout were noticed, the turnaround would be reduced to 24-hours.

As soon as the programme charter was agreed, all programme and project team members completed the baseline assessment. As new members joined, they also completed the baseline assessment. Every team member would complete the assessment once a year (or more frequently if signs of burnout appeared). Anything between 6 months to 12 months, maximum, is advisable.

Integrated wellness structure

As can be seen from Figure 11.4, the university was contracted to perform the regular assessments and advise which team members needed support, whilst EAP Solutions, who specialises in employee assistance programmes in a technical project environment, provided continuous employee wellness support and counselling services. EAP Solutions followed the principle of confidentiality. In terms of reporting, Intego management only received summarised group reports such as the one shown in Figure 11.5. Individuals whose results were problematic were approached on a one-on-one basis directly by EAP Solutions. Team members could also contact EAP Solutions of their own accord.

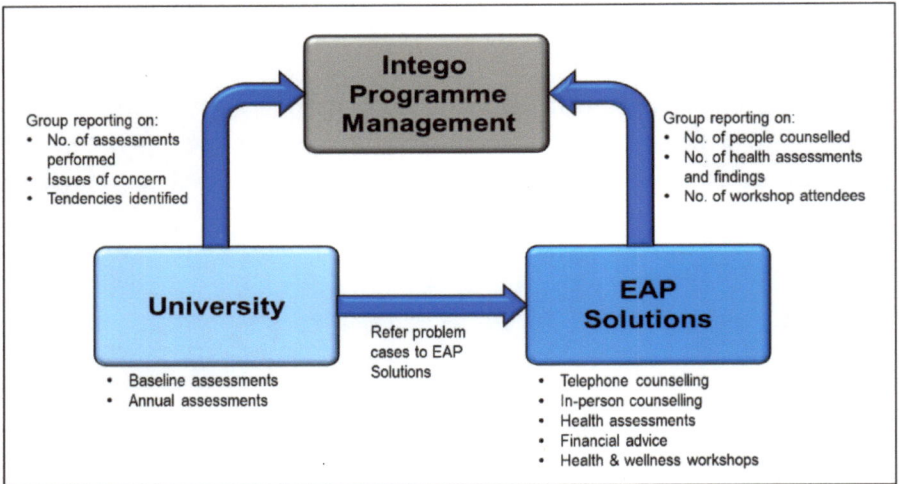

Figure 11.4: The Intego Integrated Wellness structure

Sample baseline assessments

In Figure 11.5 below, a sample baseline assessment stress report for the teams within PowerCo is depicted. The scores were done as percentages, indicating the extent to which the elements highlighted, contributed to team members' stress levels. These reports are averaged scores across teams. Only qualified members of the university had access to individual scores, which was in turn supplied to the Employee Assistance Programme (EAP) support team that was appointed on Intego.

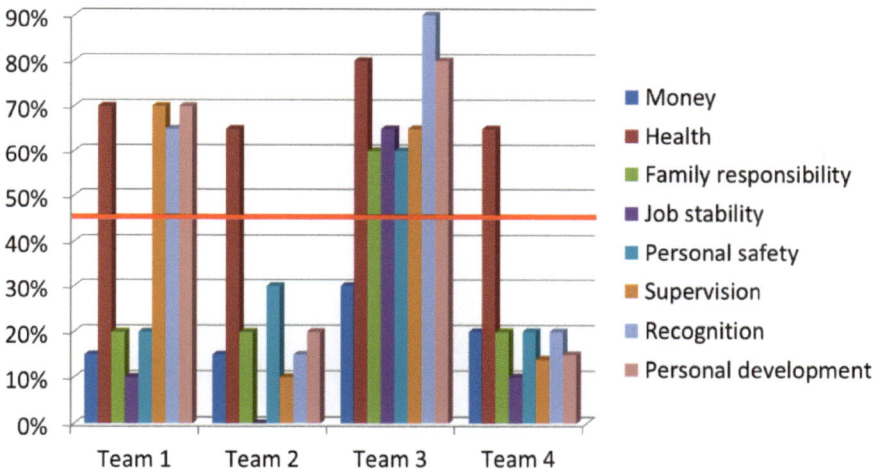

Figure 11.5: Stress Report for teams in PowerCo

From Figure 11.5 it can be seen that the four teams within PowerCo all displayed health issues which were drawn from sick leave statistics and stress related symptoms displayed.

Teams 1 and 3 also displayed:

• Concerns with supervision;

• Concerns with recognition practices, and;

• Concerns with personal development opportunities.

This provided the programme management team with a good indication of what issues to investigate and address.

To address the issues identified, Intego provided stress management courses which include healthy lifestyle coaching. Supervisors and their style of management was assessed and it soon became clear that a specific supervisor lacked critical interpersonal skills, including people skills, conflict management skills and people development skills. After doing dedicated training targeting his shortcomings, the scores improved dramatically after 6 months when the next assessment cycle was performed.

Interestingly, Team 3 displayed high scores on all elements except income. It just so happened that due to changes in the programme scope, the job security of this team was under question at the time due to possible reduction in the scope of the programme. They also lost their supervisor a month earlier and had to cope with a new manager and a new management style.

From this example, it can be seen that upon unpacking the results, one can fairly easily make certain deductions that can be tested and addressed in a practical manner.

Physical assessments

Annual cholesterol, blood pressure and blood sugar level testing are also conducted. Based on the results, each individual is given a personalised report with advice on healthy habits, suggested medical treatment, and referral to medical practitioners when required. To simplify the process and time away from work, physical and stress assessments were aligned to fall on the same day. A well-managed wellness programme can save you a lot of money and heartache, prevent schedule slip and increase productivity.

Most employee assistance programmes (EAPs) allow for telephonic and/or in-person counselling. On a programme, one should preferably have both options available. It is very important to make sure that you have a good contract in place with your EAP service provider to ensure that turnaround times meet your needs.

Practical considerations

One practical issue faced on programmes, is that in many instances some of the work that forms part of a programme is based in remote locations. In this case, one needs to assess the severity of a problem that had been identified and decide:

•   Do you return the individual back home for a few days to get the help he or she needs?

•   Do you first do a telephonic or video conference counselling session with the individual to make an informed decision?

In many cases, the stress and longevity associated with a programme put severe strain on relationships. As the programme takes on a life of its own, people's personal lives can't come to a halt. Life goes on: people have children, family members pass away, marriages go through rocky times and so forth. Senior programme managers will know that the programme and associated personnel problems becomes their 'baby'. If it screams, they have to be there. As a result, they tend to neglect other important aspects of their lives, which may even include their own health. A properly integrated wellness programme will make provision for the assessments, telephonic counselling as and when required (24 hours a day), targeted one-on-one counselling and wellness awareness campaigns and any other individual or team interventions that may be deemed necessary.

Coaching on work-life balance should be considered on all projects. Work-life balance typically includes eating habits, exercise and enough sleep. To cope with prolonged stress, these three factors can help you manage your stress and ensure your own sustainability over time. These can be done through wellness campaigns which could include intranet site information, e-mails, posters and formal or informal discussions.

The best way to approach wellness on a programme is to introduce an integrated wellness programme prior to the symptoms of poor performance, illness and disengagement manifesting.

Concluding remarks

Nurture your programme team members and celebrate your success!

In a nutshell, investing time and effort into individual and team wellness contributes to a team feeling cared for and recognised. It not only contributes to an individual's wellbeing, but also to the success, performance and productivity of the programme team. None of these activities are loose standing from achieving overall success. They are integral parts of ensuring team and programme objectives are being met.

An essential element of effective leadership (see Chapter 10) is to build a well-functioning, effective and focused team. It is always amazing to see the results that can be achieved if the leadership manage to develop an effective team. The rewards are not just marginal but can result in an order of magnitude improvement in the effectiveness and outcomes. A well-functioning team will build upon its successes and maintains a high level of performance while a dysfunctional team will spiral downwards and eventually drastic interventions will be required.

Chapter 12:
Keep Stakeholders Informed and Engaged

"The single biggest problem in communication is the illusion that it has taken place." -George Bernard Shaw

Introduction

The ultimate objective of stakeholder management is to ensure successful execution of the programme by engaging the right people that could influence or impact the success or failure of the programme at the right time.

It is all about knowing who is interested or affected by the programme, then developing and implementing a strategy to 'manage' (or interface with) stakeholders in the best interest of all. Stakeholder engagement is a continuous activity during a programme and forms the third ring in the lifecycle essentials portion of the programme management model, as shown in Figure 12.1.

Stakeholders are any individuals or groups that have a legal, financial or social interest in an organisation or programme, such as shareholders, directors, managers, suppliers, customers, government, employees and the community. Remember that 'you' are also a stakeholder in your own programme. In short, stakeholders have an interest in or concern with the programme, are affected by the programme in some manner, and could directly or indirectly have the power to influence the programme.

In this chapter we discuss stakeholder management, including the identification of stakeholders. This is followed by a discussion on a communication plan for effective stakeholder engagement and useful communication tools.

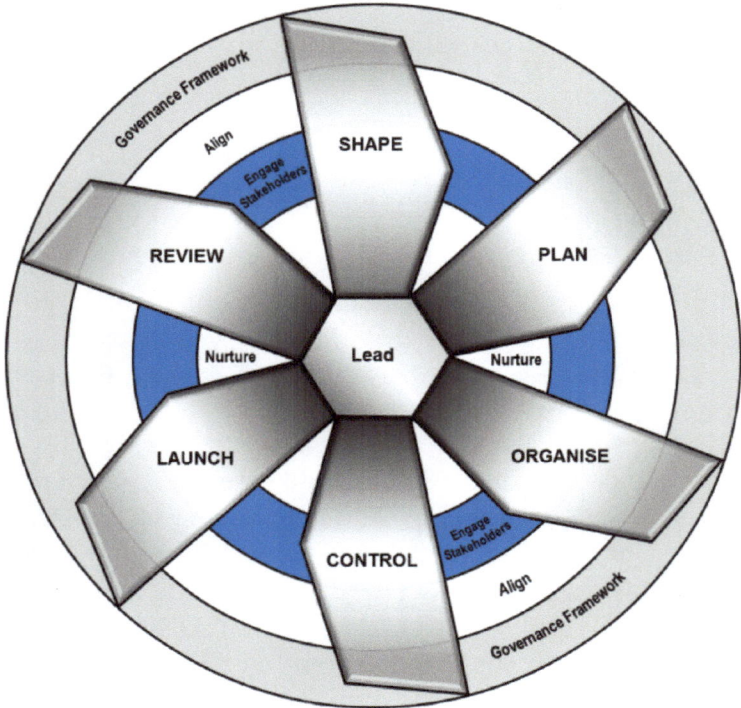

Figure 12.1: Programme lifecycle model with 'engage Stakeholders' highlighted

Stakeholder Management

Stakeholder categories

Three key stakeholder groups are of particular importance, namely company internal stakeholders, engaged contractors and external interested and affected parties, as illustrated in Figure 12.2. Each of these categories is discussed in more detail below.

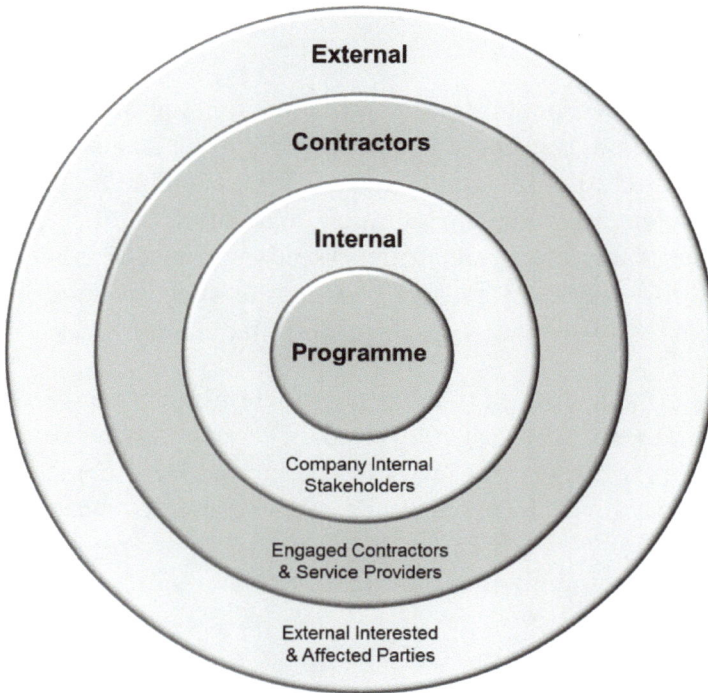

Figure 12.2: Types of stakeholders on a programme

Stakeholders have an interest in, or concern with, the programme, are affected by the programme in some manner, and could directly or indirectly have the power to influence the programme.

Company internal stakeholders include the employees and officers of the business who own the programme, any centralised company functions and other businesses within the holding company affected by the programme, as well as the programme owner team members that form part of the programme management team. Included are also those that form part of any one of the sub-project teams being managed under the programme.

Engaged contractors and service providers include those involved in executing the programme and sub-projects in all stages from design to implementation. It could include design engineers, technology suppliers, engineering consultants, managing and engineering contractors, fabricators, constructors and suppliers. It also includes all supporting contractors and consultants such as environmental consultants, business advisors, legal and financial advisors.

External interested and affected parties represent members of the local community who are interested in or affected by the project, government at the various levels (local, provincial and national, and in some instances international), non-government organisations (NGO's) or any other interest group or pressure groups, and organised labour in the form of unions that are either interested in or affected by the programme. These may also include external business competitors or potential partners.

Stakeholder management dynamic

Both the programme charter and the programme execution approach contain stakeholder management elements. When doing stakeholder identification, be sure to reflect back on these documents (refer to Chapter 4). When stakeholder engagement strategies are discussed, more examples will be explored.

Stakeholder management on a programme should never be an ad hoc activity or event. It should be a structured, yet dynamic process managed on a continuous basis throughout the lifecycle of the programme.

The stakeholder management dynamic is shown graphically in Figure 12.3. The first step is to identify stakeholders based on the three stakeholder categories discussed in the previous section. The second step is to map each stakeholder in relation to their power to influence the programme, attitude toward the programme (positive/negative) and interest level in the programme. The final step is to develop and implement a customised stakeholder engagement strategy and plan for each stakeholder or group of stakeholders, as may be appropriate.

Figure 12.3: Stakeholder management dynamic

Each of the three elements of the stakeholder management dynamic, namely, stakeholder identification, stakeholder mapping and stakeholder engagement, is now discussed in turn.

Stakeholder identification

There are specific questions that can be asked under each stakeholder category that could help trigger who the stakeholders are:

For company internal stakeholders, the following questions are relevant:

- Who provides input or feedstock?

- Who receives the output of the programme and sub-projects?

- Who can stop or give impetus to the programme?

- Who are the decision makers?

- Who can influence the outcomes of the programme?

- Who are internal resource providers?

- Who are team members?

- Who are consenters?

- Who are supporters of the programme?

- Who are opponents of the programme?

- Which areas/departments/workers are directly affected by the programme?

For engaged contractors and service providers, the following questions are relevant:

- Who could provide input?

- Who are resource providers? I.e. wellness support, legal support, or any other non-technical support

- Who are technology suppliers?

- Who are engineering consultants?

- Who are design engineers?

- Who are managing contractors?

- Who are engineering contractors?

- Who are construction contractors?

- Who are manufacturers, suppliers and fabricators?

- Who are consultants (environmental, legal, marketing and others)?

For external interested and affected parties, the following questions are relevant:

- Who provides input or feedstock?

- Who receives the output of the programme and sub-projects?

- Who can stop or block the programme or give impetus to the programme?

- Who can influence the programme?

- Who are consenters, e.g. for environmental permitting?

- Who are resource providers?

- Who are beneficiaries in the communities or government?

- Who in the external environment could be supporters or opponents?

- Who are vulnerable groups that could be affected by the programme?

- Which central, provincial or local government departments are involved?

- Which NGO's are affected?

- Which community or local pressure groups could involve themselves?

We return to the Intego programme to illustrate how these principles are applied in practice. For the sake of brevity, only the internal stakeholder identification will be shown.

Intego Environmental Compliance Programme (IECP)

Internal Stakeholder Identification

The list of internal stakeholders was compiled using the list of trigger questions shown above. The outcome is reflected as Table 12.1.

Table 12.1: List of internal stakeholders

Question	Stakeholder
Who provides input or feedstock?	• Intego Environmental Department • MiningCo • Gas and Coal Operations
Who receives the output of the programme?	• LogisticsCo • UtilityCo
Who can stop the programme?	• The IHL Board (approval body) • Programme Sponsor
Who are the decision makers?	• The IHL Board (approval body) • Programme Sponsor
Who can influence the outcome of the programme?	• Intego Holding Ltd • MiningCo • PowerCo • Pipeline Pro • GridCo • Logistics Pro • UtilityCo.

Question	Stakeholder
Who are the internal resource providers?	• Pipeline Pro • Logistics Pro • GridCo
Who are programme team members?	• Programme Sponsor • Sponsor of gas projects • Sponsor of coal projects • Programme management team • Project management teams, • PowerCo operations and maintenance teams at each power station
Who are consenters of the programme?	• MiningCo • PowerCo • Pipeline Pro • GridCo • Logistics Pro • UtilityCo
Who are supporters of the programme?	• UtilityCo
Who are opponents of the programme?	• PowerCo

Stakeholder mapping

Stakeholder mapping can only be done once stakeholders have been identified. Unfamiliar stakeholders may require further analysis and research. It is recommended to have at least an indication of the motivation, connections and power of the different stakeholders.

It is now necessary to map all stakeholders in terms of their interest in the programme and their power to influence the outcome. West (2010) finds that the matrix in Figure 12.4 can be effectively used to map your stakeholders. Stakeholders listed as having a low interest in the programme, and have little power to influence the outcome of the programme, require a monitoring strategy, whereas those with a high interest and high power to influence need to be managed closely.

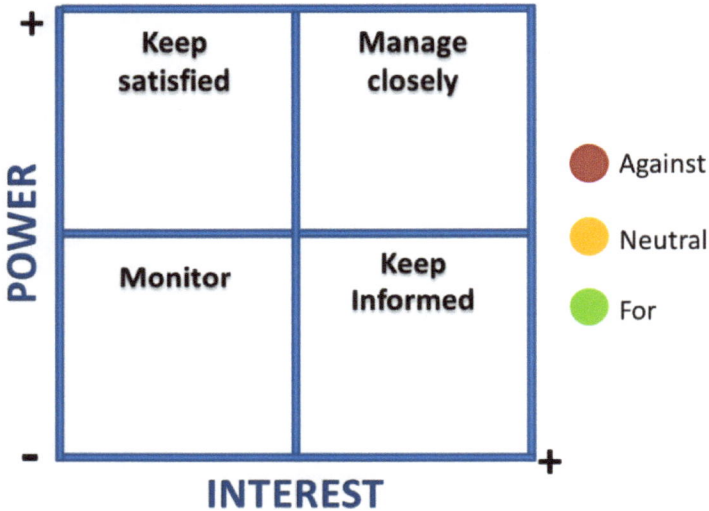

Figure 12.4: **Stakeholder mapping matrix** (adapted from West, 2010)

Stakeholder positions on the matrix can change during the lifecycle of a programme, hence project team members should keep a careful eye on how stakeholders are positioning themselves in relation to the programme. This will allow timeous changes to be made in the approach to these stakeholders.

Should a natural group of stakeholders hold the same view, you can cluster them together, but you still need to define who your entry point will be and how you will engage the stakeholder group. However, should any given individual hold a very firm view, either positive or negative, it is a good strategy to engage him or her on an individual basis.

Because stakeholder maps can become cluttered, it is recommended that you develop three separate maps, one for company internal stakeholders, one for engaged contractors and one for external interested and affected parties. Developing strategies on how to engage these stakeholders can then also be based on the three categories.

It is recommended that you develop three separate stakeholder maps, one for company internal stakeholders, one for engaged contractors and one for external interested and affected parties. This will ease the development of different approaches with the different categories of stakeholder.

We return to the Intego case study to illustrate the handling of two internal stakeholders who hold very strong opinions on the programme.

Intego Environmental Compliance Programme (IECP)

Company internal stakeholders and stakeholder map

The key players (groups and individuals) from the internal stakeholder identification process as described in the previous section on stakeholder identification are:

- **Intego Holdings Limited (IHL):** The overarching holding company responsible for the overall value chain, and initiator of the programme. Keeping this decision maker informed and abreast of progress is critical, as is managing their expectations;

- **MiningCo:** MiningCo is a fully owned subsidiary of IHL and consists of two divisions, namely gas operations and coal operations. Gas operations comprise the gas field operations and gas conditioning facilities and the coal operations comprise the mining activity and coal classification. Changes to the unit operations to reduce environmental omissions impacts their processes as well, and this implies that they are a critical stakeholder that needs to be kept in the loop at all times in terms of how the project impacts their product specifications and processes;

- **PowerCo:** PowerCo is a wholly owned subsidiary of IHL and comprises the power generation facilities of Intego, be it gas-fired or coal-fired power stations. They are seriously affected by the programme in that their operations would require major revamps to be able to meet the new environmental specifications. Hence they are also opposing the pro-active changes by the organisation and would rather postpone some of these changes;

- **Pipeline Pro:** This is a joint venture company between Intego and external parties, responsible for the maintenance and operation of a network of pipelines for the transfer of gas to consumers. A major gas consumer is Intego's gas-fired power stations. The changes will add more pressure to the pipeline capacity, as the demand for gas as a feedstock is increased. Thus another major player and heavily affected by the changes;

- **GridCo:** This is a joint venture company of Intego and external parties for the maintenance and operation of an electrical distribution network. GridCo is affected by the changes in that they need to be aware of the activities throughout the programme so as not to affect the stability of power supply;

- **Logistics Pro:** This is a non-affiliated service provider charged with the transport of coal feedstock to the coal-fired power stations. Logistics Pro will be affected in the sense that coal demand will decrease and may lead to job losses over the long term;

- **UtilityCo:** This is a non-affiliated service provider charged with the sale of electrical power to consumers. The changes are likely to result in an increased cost of power which UtilityCo will have to manage carefully;

- **Intego Environmental Department:** The corporate Intego Environmental Department is in strong support of the programme because of the improvement in environmental performance of the organisation. They believe that the programme will extend Intego's licence-to-operate, and;

- **Owner programme and project teams:** The teams responsible for executing the programme and sub-projects. Due to the integrated nature of a programme, the various sub-project teams need to be aware of one another's progress.

Apart from the above companies and groups, two individuals were also listed due to their strong opinions. They are:

- **Jane Perkins**, a member of the IHL Board, is a very strong supporter of the programme and was one of the initiators, and;

- **John Jones,** Managing Director of PowerCo, is very outspoken about the programme and upset that he is "held at ransom" to make the changes to his business.

All the internal stakeholders are plotted on the populated internal stakeholder map in Figure 12.5. Apart from the company groups, you can see that Jane Perkins and John Jones were listed as individuals.

Colour coding is used to indicate whether the groups or individuals are for (green), against (red) or neutral (orange) towards the IECP. The more red dots in the top right-hand quadrant of the matrix, the bigger the stakeholder management task is. Red dot stakeholders have the power to stop or delay the programme.

Note that a similar exercise is required for the engaged contractors and service providers, as well as the external interested and affected parties.

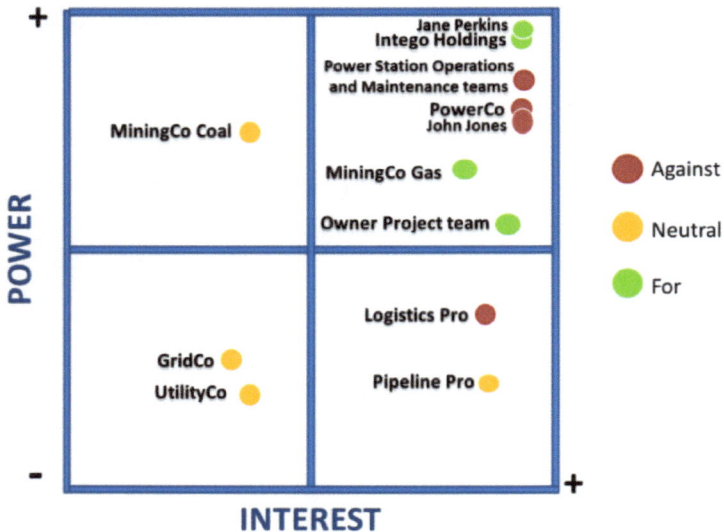

Figure 12.5: Intego Internal Stakeholder Map

Stakeholder engagement

Considering the stakeholder map in Figure 12.5, the stakeholders in the right top quadrant of the matrix must be actively managed. Due to their high level of interest as well as power over the programme, they could either support or block the programme's progress, depending on how they are disposed towards the programme. Supporters should be engaged to the extent that they remain supporters, whilst those opposing the programme should be actively engaged to keep them up to date with the programme, but also to provide them with enough information to positively impact their opinions. A good strategy is to firstly engage those that are strong opinion makers or decision makers. Opinion makers can be used to
2

Engagement with stakeholders can take place using the following engagement platforms, depending on the purpose of the engagement:

- **One-on-one meetings:** These are useful for sharing information and engaging a person on a personal level. It could help in building relationships;

- **Personal interviews:** These can be used to gather information on issues of concern;

- **Presentations:** Presentations to larger audiences are effective for communication in one direction;

- **Workshops:** Workshops are the best option if you wish to really engage stakeholders and get their input and buy-in as a group;

- **Focus groups:** These are useful in solving specific problems or addressing specific concerns. A focus group is a smaller group addressing a very specific issue;

- **Committee meetings:** Small groups of stakeholders reviewing progress with the intent to make or support decisions, or to review progress and plan activities;

- **Internal road shows:** Internal road shows typically comprise displays and presentations and are a very effective way of doing mass communication;

- **Public meetings:** Similarly public or town hall meetings are used to communicate to large groups of external community stakeholders;

- **Think tanks:** Think tanks are groups of experts getting together to solve specific complex problems;

- **Surveys:** Surveys could be used in many ways such as opinion surveys, suggestion boxes and feedback forms in order to collect qualitative or quantitative data;

- **Advisory committees:** Advisory committees play a key role in programmes and could be done using either internal or independent external advisors. These could be done on a regular or an ad hoc basis;

- **Collaboration platforms:** These are typically done on-line using collaboration technology, mostly used to share ideas, questions and best practices with groups of people with similar interests;

- **Forums:** Forums are a useful way of getting like-minded people together and keeping them up to date with latest developments in industry and the organisation;

- **Portals:** Intranet portals are typically used for internal stakeholders and internet portals for external stakeholders;

- **Bulk SMS/text portals:** Text groups are compiled for bulk communication of brief messages, and;

- **Hotlines:** Hotlines can be set up for stakeholders to seek advice or report grievances.

The above list merely contains some examples of engagement platforms that could be used on programmes to ensure that sound one- and two-way communication and engagement takes place. More options are available and even more will become available in future.

A communication and engagement plan must be developed in support of the stakeholder map. Figure 12.6 is a graphic model to consider in developing your communication and engagement plan.

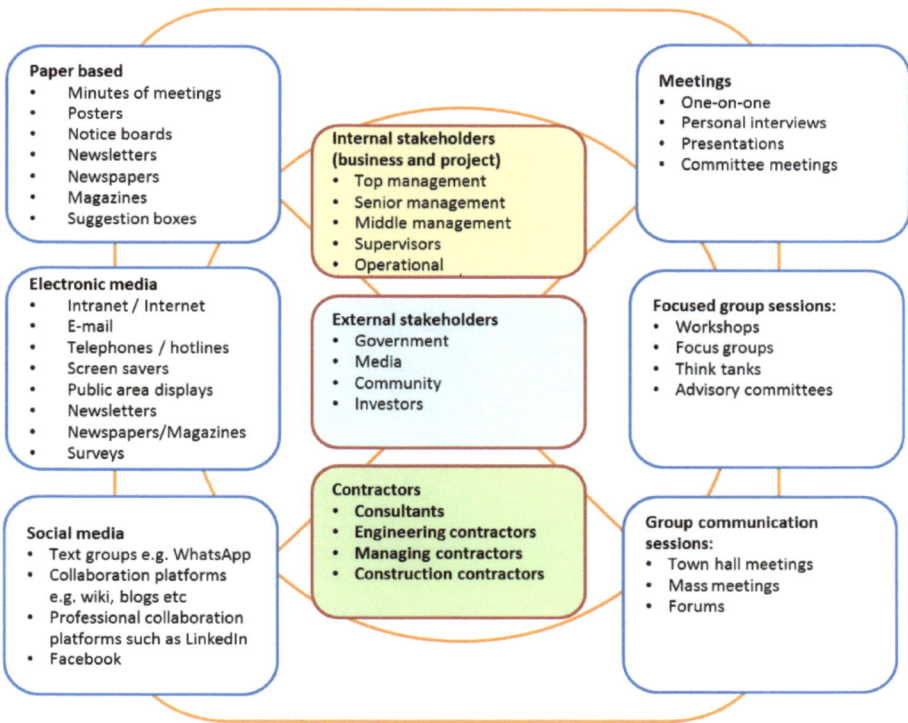

Paper based
- Minutes of meetings
- Posters
- Notice boards
- Newsletters
- Newspapers
- Magazines
- Suggestion boxes

Electronic media
- Intranet / Internet
- E-mail
- Telephones / hotlines
- Screen savers
- Public area displays
- Newsletters
- Newspapers/Magazines
- Surveys

Social media
- Text groups e.g. WhatsApp
- Collaboration platforms e.g. wiki, blogs etc
- Professional collaboration platforms such as LinkedIn
- Facebook

Internal stakeholders (business and project)
- Top management
- Senior management
- Middle management
- Supervisors
- Operational

External stakeholders
- Government
- Media
- Community
- Investors

Contractors
- **Consultants**
- **Engineering contractors**
- **Managing contractors**
- **Construction contractors**

Meetings
- One-on-one
- Personal interviews
- Presentations
- Committee meetings

Focused group sessions:
- Workshops
- Focus groups
- Think tanks
- Advisory committees

Group communication sessions:
- Town hall meetings
- Mass meetings
- Forums

Figure 12.6: A communication and engagement model

The various engagement vehicles shown in Figure 12.6 should be considered in line with your stakeholders and their needs. Carefully consider frequency of engagement to ensure stakeholders are kept abreast of what they need to know, but also to avoid overloading them in which case they will become disinterested or irritated.

Again focussing only on the internal stakeholders, an extract from the Intego communication and engagement plan is presented below.

Intego Environmental Compliance Programme (IECP)

Communication and Engagement Plan

An extract from the Intego Communication and Engagement Plan, covering the company internal stakeholders is given in Table 12.2.

Table 12.2: Extract from a Communication Plan

Engagement Platform	Frequency	Distribution/Include these stakeholders	Originator of Document/ Organiser of Meeting	Chairperson
Programme Steering Committee meetings	Quarterly	Power Station Operations Manager Power Station Maint. Manager PowerCo Operations Manager MiningCo Gas Operations Manager Logistics Pro Operations Manager Pipeline Pro Operations Manager GridCo Operations Manager UtilityCo Business Manager MiningCo coal Mining Manager Programme & project specialists	Programme Director	Sponsor
Steering Committee minutes	Monthly	All MDs, Business and Operations Managers of the business units reflected on the internal stakeholder map	Programme Director	Not applicable

Engagement Platform	Frequency	Distribution/Include these stakeholders	Originator of Document/ Organiser of Meeting	Chairperson
Master communication packs	Quarterly update	All new recruits for the internal programme team	Communication Manager	Not applicable
Programme portal	Monthly update	All internal stakeholders	Communication Manager	Not applicable
Organisation charts and contact lists	Monthly update	All internal stakeholders	Project Administrator	Not applicable
Monthly one-on-one discussions	Monthly	Jane Perkins John Jones Power Station Operations Manager Power Station Maint. Manager	Sponsor	Sponsor

Note in Table 12.2 above that it is also very important to ensure that clear accountability is assigned to activities.

As a rule-of-thumb in terms of how to engage different stakeholders, the stakeholders low on interest and power can generally be engaged in groups via written communication; see Figure 12.7. Stakeholders high on interest and power should preferably be engaged on a one-on-one basis.

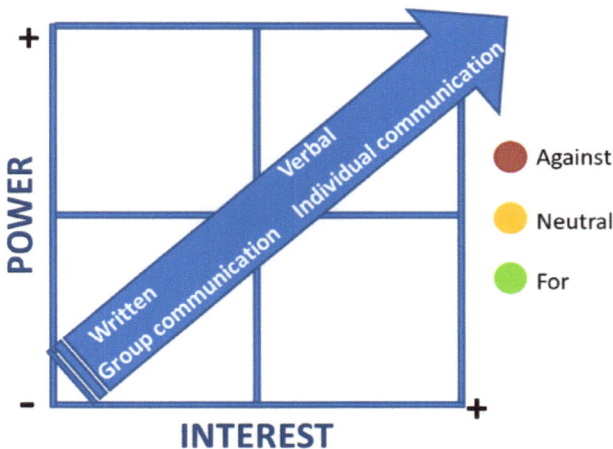

Figure 12.7: Generic engagement guideline

Stakeholders can move to other quadrants

Stakeholders will not necessarily remain in the quadrant where they were originally plotted, throughout the programme. Moves can be triggered by actions taken by the stakeholders themselves or by specific interventions from the programme communication team.

On the Intego programme, there was a single community member, a certain Mr Hugh, who was against the programme due to the potential job losses. He actually submitted a written complaint to Intego. He approached other NGOs and community pressure groups and soon gained enough support to also draw the attention of the media and unions. Over a very short period of time, he moved up into a position of power due to his efforts to solicit support. This situation is illustrated in Figure 12.8.

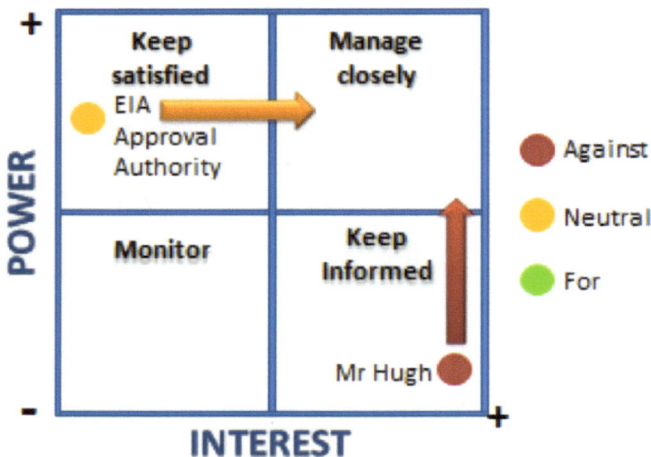

Figure 12.8: Examples of stakeholder strategies in Intego

In managing stakeholders, one has to make a conscious decision if you wish to engage with a stakeholder with the objective of trying to prevent them from moving into a different quadrant, or if you wish to try and influence them to become for example either more interested or more positive.

Another example was the approval authority for the environmental impact assessment (EIA). They had huge power in terms of approval of the EIA, but showed little interest. This meant that the application remained in their inbox and they had no sense of urgency to get it processed. A strategy was then developed to capture their interest and thus to move them to the top right quadrant. This consisted of directly engaging with them and providing the information in such a way as to capture their attention, especially focusing on the environmental benefits associated with the programme. After repeated personal contact with them, they developed an interest in the programme and the necessary approvals were processed smoothly.

The success of stakeholder management is first of all knowing who the stakeholders are and how they are positioned, and secondly making a conscious effort in terms of engaging them in such a way that they have an appropriate level of interest in the programme and that their attitude towards the programme does not harm the programme in any way. This can only be done by pro-active, well-planned and continuous engagement.

Programme Communication

Introduction

Over and above engagement strategies that get devised on a case-by-case basis, there are certain 'must do' engagement efforts on a programme. In this section of the chapter, some examples of what was done on the Intego programme are reviewed.

As was said before, stakeholder engagement on a programme is about knowing who your stakeholders are and understanding how they can contribute to the success or failure of your programme. This provides the basis for your strategies to keep stakeholders aligned and informed in order to solicit the right levels of engagement.

Communication on any programme is the sum total of keeping the whole programme team and key stakeholders abreast of what is happening on the programme, and why! Communication on a programme has various phases. The first phase is to set up the basis for communication. Thereafter all communication is about capturing the changes that happen during the lifecycle of a programme, why changes are happening, and the impact thereof. Good communication should always cover the what, why, when, where, who and how aspects.

Master Communication Pack

A challenge on programmes is the extended lifespan thereof and somehow key information always gets lost along the way. A project is initiated within a certain context with very specific assumptions and objectives, but these can change over time. The challenge is to keep a consistent storyline together from a communication perspective, whereby the programme team always has an up to date communication pack that can be used to orientate new members on the project. The up to date storyline is referred to as the master communication pack

The master communication pack helps programme and project members to ensure consistent communication. In Chapter 4, the importance of the programme charter was discussed. The charter is the starting point of the master communication pack. Tracking changes to the charter, key assumptions and decisions underlying the programme and keeping the communication pack up to date minimises the risk that every person draws up their own slides, which represents their own translation of the message. A master communication pack should be coordinated by the programme administrator and published to the required stakeholders each time it gets updated.

A typical master communication pack contains items such as:

- Business objectives with its underlying assumptions (charter: why you doing the programme);

- Project economic drivers;

- High level programme schedule/framework;

- Programme objectives – scope, quality objectives, time line (framework plan), cost of the programme;

- The safety, health & environmental (SHE) objectives & philosophy;

- Any philosophies relevant to the programme phase at hand;

- High level scope (in the form of the work breakdown structure of the respective projects making up the programme);

- High level schedule of the active phase;

- Execution strategy - how the projects will be managed within the framework set out in the charter;

- Programme organisation structure for owner team and contractors;

- Key interfaces on the programme and sub-projects that should be managed;

- The approach/philosophy with regards to Broad Based Black Economic Empowerment (BBBEE), which is specific to South Africa;

- The approach to risk management with a summary of key risks on the risk matrix;

- Recognition and reward system;

- Site information e.g. accommodation, permits, etc.;

- Maps of the area around the site;

- Photographs showing progress, and;

- Any relevant information that the programme team may need in the particular phase of the project.

The master communication pack can be supplemented with a list of frequently asked questions (FAQ's), which is always a useful tool for capturing answers to questions that are foremost on people's minds. The master communication pack can very effectively be used for the orientation of new team members.

The above is merely a guiding list. Each client has their specific needs and expectations. That said, the above should be regarded as a minimum requirement for a master communication pack. Each of the items listed is supported by a detailed document.

⚠️

The master communication pack is a concise summary of the programme principles. It is used to communicate to key stakeholders and, in particular, the team members as they get mobilised on the programme during its lifecycle. It is also very useful for communication to external strategic stakeholders.

One of the major aspects that changes more often than not, is the programme organisation structure. The structure typically changes as specific project teams are mobilised or demobilised. Changes in scope can also lead to additional or reduced resources. Due to the fluid nature of a programme, projects may be envisioned which could later on be considered as non-viable in the bigger context. For this reason, programme teams tend to gear up for projects they envision to kick off, and if they do not realise, for whatever reason, they downsize.

It needs to be emphasized that the master communication pack is not a detail presentation that unpacks every project on your programme. It is rather a strategic communication pack which provides the bigger picture and how the pieces fit into that picture.

Portals

A useful tool on programmes is a portal where key information can be made available to the programme and project teams and other stakeholders.

Two types of portals can be established, namely an internal portal and an external portal. Internal or intranet portals are aimed at giving the project team quick and effective access to project information. This would include vendor's lists, designs, cost & schedule reports, internal contacts, framework plans, schedules, budgets, etc. The only purpose of an intranet portal is to create quick and easy access to information for team members via a user-friendly interface. A separate internet page with high level information can be created for the purposes of communication to an external audience (external to the team and or company). Information should never be duplicated on systems: Keep the information in the system where it belongs, and create links from your portal to there.

The portal would typically reflect your programme structure at the highest level. It is recommended that a portal structure be set up according to your programme phases, rather than the programme modules. There are two reasons for this recommendation: Firstly, in any given programme, there is a lot of common information that will be duplicated for the different modules, and secondly, the modules are likely to change over the lifespan of the programme.

The use of an intranet site for information sharing for the Intego Programme is discussed below.

Intego Environmental Compliance Programme (IECP)

Programme Intranet Layout

The front page of the IECP intranet site is shown in Figure 12.9.

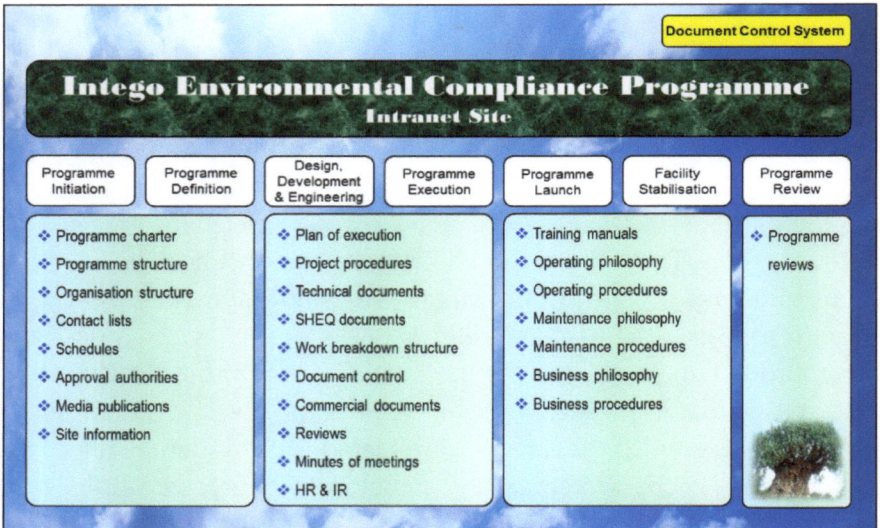

Figure 12.9: Intego Intranet front page

By clicking on any one of the links, access is granted to the documents described in the title. This can be anything from the charter, the programme structure, schedules, approval authorities and any media publications.

In the example, several of the phases are combined because the relevant documents are the same for the combined phases, but the level of detail increases as you move through the programme.

Let's discuss the drop downs one by one:

Programme Initiation and Programme Definition phases

During the programme initiation and programme definition phases, baseline information is developed and continually updated. It is important to create access to the latest version of these documents.

- **Programme charter:** All revisions of the programme charter should be accessible from the portal. This allows the team to refer back to older revisions in the event that changes may have occurred;

- **Programme structure:** The programme structure would reveal the different modules/projects within the programme. The intent with the programme structure is to provide the bigger picture;

- **Schedules:** This section would supply all schedules. The overall programme master schedule, as well as the summary schedules for each project/module, and detail schedules within each project. The structure should provide for a layer of schedules. These schedules are typically stored on a project management system, and will merely be accessed from the portal via a link. This would also address confidentiality issues through access control;

- **Approval authorities:** Due to the complexity of programmes, their approval authorities also tend to be quite complex. It should be made available via the portal and should always be kept up to date through proper version control;

- **Media publications:** All internal and external publications on the programme should be posted under Media Publications on your intranet site, and;

- **Site information:** Very important information for any team member, and especially new team members, is site information. This includes maps to and from different sites that are relevant to the programme. It also includes building layout plans indicating who sits where, emergency exits, canteens etc.

Many programmes tend to have sites in remote locations. Useful information will include information on the towns in which the programme is active. This would include information on accommodation, entertainment and medical facilities.

Design, Development, Engineering and Programme Execution

- **Plan of Execution:** All versions of the execution plan for the programme should be stored here. As the programme progresses and is likely to change, the updated programme plan of execution should be made available to the team via your portal;

- **Project Procedures:** These would include all programme and next level project specific procedures, including quality;

- **Technical Documents:** Relevant technical documentation which the teams use on a regular basis should be linked to the intranet site and be easily accessible. Care should be taken to have the latest version of any document available;

- **SHEQ:** This could include safety, health, environmental and quality philosophies, procedures, EIA documentation, audits and progress reports;

- **Risk Register:** This covers the complete programme risk register and risk analysis with its associated action plans and progress reports;

- **Work Breakdown Structure (WBS):** The Work Breakdown Structure is one of the most important documents on any programme or project. The current WBS and previous versions should be accessible;

- **Document Control:** This would provide you a link to the system your company uses for document control;

- **Commercial Documents:** A major pitfall is that many companies either have poor commercial document control, or they keep it separate from the project document control. Whatever your setup, ensure that you have a sound document control system for all commercial documents from day 1, and that all updates and changes are effectively tracked and managed. Access to these documents via a single interface point is crucial;

- **Reviews:** Programme and project reviews should occur at least every quarter and copies of the reviews should be accessible from the portal;

- **Minutes of Meetings:** All minutes of key meetings should be maintained and easily accessible, and;

- **Human Resources (HR)/Industrial Relations (IR):** HR and IR are critical on any project. This would typically include documents such as organograms, resource plans for the different phases of the programme, programme team contact lists, recognition award notifications, relevant programme HR & IR policies and procedures, HR & IR templates such as leave forms or acting documents, or any other relevant HR & IR documentation.

The biggest challenge is not to populate the above mentioned documents, but to keep them updated. A dedicated programme administrator is required, who works for the programme director and who, on a continual basis, updates the organograms and team contact lists.

Programme Launch & Stabilisation

At this stage of the programme lifecycle, operating philosophies and procedures, maintenance philosophies and procedures, as well as business procedures and training manuals should be complete and accessible. During the earlier phases of the programme lifecycle, the initial revisions of these manuals and procedures will be published under these icons.

Programme Review

Last but not least, the programme and each sub-project should be reviewed at the end of the programme lifecycle to establish if it reached its original intent as stipulated in the programme and project charters.

Document control system

At the right top of the portal a link has been created to the programme document control system. Each programme always has its own extensive document control system where all documents are saved and tracked in a very methodical manner. Experts who work on these systems must find it easy to access information, seeing that they work on the detail level that is familiar to them.

The major benefit of a portal is easy and simple access to information already stored on the document control system. From a management point of view, a portal provides the bigger picture with quick access to different levels in your document control system. It is very important not to duplicate information on different systems to reduce the complexity of updating several sets of documents. The portal is thus merely an intranet interface with links to where the information is stored on storage systems such as SharePoint.

The client retains accountability for only a portion of the information and the rest is retained by the contractors who normally operate on different systems. There has to be a process in place where documents are transmitted from contractors' systems to the owner organisation's system.

It is considered essential to have a document control system in place which enables documents to be transmitted from contractors' systems to the owner organisation's system.

Organisational charts and contact lists

The two items that add significant value when it comes to communication are the programme organisational charts and the programme contact list.

Contractors generally need to be very well organised in terms of their contact lists and organisation charts. They need to account for the resources they use on the programme or sub-projects. The added complexity for the owner organisation's employees is that not all team members work full-time on the programme. Yet, they need to be included in communication on the project. Without a complete list of owner team members, full-time and part-time, as well as external resources co-opted by the owner, communication becomes very difficult.

The challenge is to keep the information up to date throughout the programme.

Having access to the organisation charts and contact details of fellow team members, saves time and effort on trying to find out who to talk to and getting hold of their contact details.

Distribution lists

From the contractor and owner organisation contact lists, one is now able to develop standard e-mail distribution lists. The lists would be available on the team's e-mail system address book.

The typical occurrence is that project managers and owner representatives spend a lot of valuable time to try and find out who is who and where to get a hold of them. On smaller projects this is not such a major issue, but working on a programme or a mega project where at any given time you could have hundreds to thousands of people working on the project, having good information is critical.

The programme management team and the project sub-teams should at least have e-mail distribution lists. These distribution lists will be set up for the programme and be updated on a continual basis. Once again, the benefit is time saved.

Meeting analysis

An intervention that can save time is a programme meeting analysis. This entails drawing up a visual chart of all meetings on strategic, tactical, and operational level. Each meeting needs to have a defined chairman, scribe, purpose, frequency and list of attendees.

Mapping what information or decisions should flow from one meeting to another helps in scheduling meetings in the right sequence. With the aid of this map, one can easily start questioning why certain people are attending certain meetings, or challenge why several different meetings address the same issues.

Meetings take up a large amount of time and can often be unproductive. Through this intervention, meetings can be streamlined and thereby free up time of key resources.

Other communications tools

Due to the exciting technological age we live in, there are many creative tools that can be used to communicate critical information, and to motivate and excite teams. Two favourites are screen savers and public area displays.

Screen savers can be very effectively used to communicate concise messages. Software is freely available whereby you can manage the screen savers of all the computers in a group. Safety campaigns, wellness messages, programme progress and much more can be posted on the screen savers from a single computer and seamlessly appear on all the team members' computers.

A second, equally exciting, tool is public area displays. The drawback of screen savers is that it only reaches employees with computers. To cater for employees who do not have computers, large flat screens can be set up in break areas or tea rooms, where the same messages are displayed, and perhaps more. The programme team is always targeted, but if a programme affects an operation or plant, it is advisable that the plant personnel also be included in such an initiative.

Experience has shown that one should not let any message run on a screen saver for longer than three days, messages should be concise, and standardised branded formats should be used. The success of such a system lies in good administration and continuous updating

We return to the Intego case study to show how screen savers and public display areas are used for maximum communication impact.

Intego Environmental Compliance Programme (IECP)

Use of Screen Savers and Public Area Displays

Safety achievements, recognition messages, general communication, safety awareness messages, wellness awareness messages and high level progress reports are posted to the teams via screen savers and public area displays. An example of such a posting is shown in Figure 12.10.

It proved to be an extremely effective tool that has an immediate impact. What works exceptionally well is to publish photos of team members on the screen savers, either where they are celebrating success or working safely. It creates excitement in the team.

Catch people doing things right and share widely...

Figure 12.10: Example of a screen saver

Programme team mass gatherings

Internal communication requires a hands-on approach with daily involvement in the programme to ensure team members are up to date with what is happening on the programme. Relevant and effective communication mechanisms must be put into place and managed.

It is advised that the programme team should have a mass gathering at least once a year where management reiterates their objectives and the direction of the programme. They can also use this opportunity to share any changes in direction. What works well and tends to be cost-effective is a breakfast gathering at a large conference facility. Try and involve as many people as possible on the programme to ensure everyone is one the same page. Avoid strategic communication via supervisor levels. Supervisors do not necessarily see the bigger picture and may only communicate their perspective. In leading up to such a mass communication gathering, make sure that the members of your programme leadership team are aligned. Remember, programmes involve different business units where each have their own objectives and targets which are not always that well aligned. Make sure to have a well-planned agenda and a dry-run with key stakeholders prior to entering into a mass communication session.

Stakeholder management resources

Even though our focus has been on internal stakeholder engagement, it is worthwhile to consider one aspect of external engagement. With programmes of the nature of the Intego programme, where legislation plays a significant role, pro-active and continual engagement with government is paramount. It would ensure that the correct targets are being pursued in light of the fact that this programme is pro-actively addressing the impact of changing legislation on the business and its licence-to-operate.

Most companies have a government relations officer who formally and informally engages with government on different levels. Whatever your company structure is, remember to follow the correct channels. Only the person who is empowered to engage with government should do so. From a strategic perspective companies interface with government on multiple levels and should at all times ensure that one voice and one message is heard by government.

More often than not, programmes tend to be very political in nature. In the case of Intego, the ultimate product is power supply. Power supply being a key resource, makes the programme even more political in nature. Due to this tendency, it is strongly advised that a programme is supported by a professional external communication expert. The services of such an individual should be available readily throughout the lifespan of the programme. He or she should at all times be up to date with any changes to the charter and the programme schedule and the possible consequences of changes. This will enable the external communications expert to ensure sound communication to the market to avoid adverse political consequences. The external communication expert thus functions on a strategic and political level on a programme and is typically a person employed by the owner company in the corporate stakeholder communication department.

A full-time government relations consultant was appointed for the Intego programme that is used to facilitate all interaction with government departments. He continually interacts with the Minister of Minerals and Resources and the Minister of Water and Environmental Affairs to ensure that the targets being set and actions being taken are in line with governments' vision and strategy.

Complex stakeholder engagement dictates having dedicated stakeholder engagement resources as part of the programme team; appoint them earlier rather than later.

Concluding remarks

Stakeholder identification, mapping and management is a very formal process. Care should be taken to identify stakeholders that could impact the success of the project and it should be planned how they will be engaged throughout the lifecycle of the programme. If full time resources are required to help in the process, mobilise them as early as possible, but only when it makes sense. Communication is a key mechanism used during stakeholder engagement. It should be used in a fit-for-purpose manner.

In the next chapter, we address the very important matter of the alignment process.

Chapter 13:
Keeping the Team Aligned

"Effective leaders help others to understand the necessity of change and to accept a common vision of the desired outcome." - John Kotter

Introduction

Alignment is a continuous process throughout the lifecycle of any programme supported by leadership, nurturing and stakeholder engagement activities. It forms the third ring of the lifecycle essentials of our programme management model, as depicted in Figure 13.1. In effect, these lifecycle essentials form the glue that keeps it all together during programme shaping, planning, organising, controlling, launching and review.

Over and above the stakeholder management activities described in the previous chapter, the parties actively involved in developing and implementing the programme need special attention in order to ensure they all work continuously towards the same objectives. In this chapter the approach to ensure alignment, both amongst company internal parties as well as with the contractors involved in the programme, are discussed.

Alignment in Context

Definition

The Construction Industry Institute (CII, 2014) defines alignment as "the condition where appropriate project participants are working within acceptable tolerances to develop and meet a uniformly defined and understood set of project objectives". Following an alignment process thus ensures that there is a common understanding of the owner's business and project objectives, as well as the owner management team's expectations.

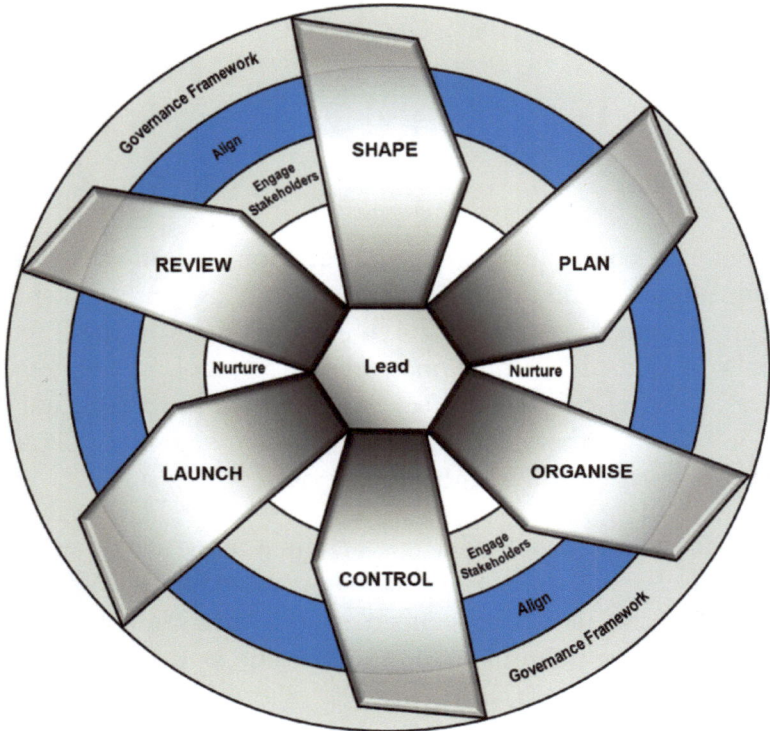

Figure 13.1: Programme management model with 'Align' highlighted

This allows for focused effort by all teams in support of these objectives.

Alignment is not an action or single step to be taken during programme execution, but rather a process. The alignment process entails two key steps, namely framing and team alignment.

Framing

Framing is the process during which the project boundaries such as the business definition of victory, the cost, time and quality requirements as well as governance and other business directives are discussed amongst the key decision makers within the business and agreement is reached on each item. Interactivity during the framing discussions and reaching consensus on the way forward is essential to concluding a successful project framework. When planning a framing workshop, one should endeavour to ensure participation by all and to prevent the session from becoming a one-way presentation.

Any project or programme is in essence a structured change and the framing meeting is focused on setting in motion a process for successful and sustainable change. Sustainable change requires a clear vision of the end-state (business objectives), as well as an understanding of the present situation; the difference between the two being the gap. With this information, the reason for the change can then be explained and 'sold' to stakeholders. The above on its own is, however, not sufficient as without a clear understanding of a process to move from the present to the future state, the implementers and stakeholders will experience significant anxiety and frustration in trying to set the process in motion. All three of these elements (clear vision, appreciation of the present state and gap, and an agreed way forward) are therefore addressed during the framing session.

The agenda for a framing meeting should cover the high level principles and guidelines (what must be achieved) and not the detail of how to actually do the work. Attendees to a framing session should be the key decision makers involved in shaping the opportunity. In order for the attendees to focus on the content provided during such a workshop, it is normally advantageous to use an experienced facilitator to manage the framing process, coordinate the discussions, ensure input by all and ensure that agreement is reached on each topic under discussion.

The framing meeting firstly elucidates the overall vision and business objective for the programme and ultimately culminates in the programme charter. It also includes agreement on a high-level roadmap, killer risks and major assumptions underlying the programme objectives. The pro-forma list below details the items normally covered (depending on the programme context) during the first session of the framing meeting (for details on the content of each element refer to Chapter 4 on initiation and shaping).

Pro-forma contents for first framing session

1.   *Business overview*

1.1   *Background and introduction*

1.2   *Overall vision*

1.3   *Business objective(s) and key characteristics thereof*

1.4   *Programme scope*

1.5   *Boundary conditions and constraints within which the team must operate*

1.6   *Major assumptions underlying the vision and objectives*

1.7   *Potential influences*

1.8   *Killer concerns or risks*

1.9   *Exit indicators or abandonment criteria*

1.10   *Key stakeholders to engage*

1.11   *Community relations and corporate social investment*

1.12   *Legislative and governmental SH&E requirements*

1.13   *Sponsorship*

Once the business overview has been completed, the second part of a framing workshop now requires the development and agreement of the high level plan going forward.

Pro-forma contents for second framing session

2.   *The way forward*

2.1   *Programme management principles to be followed*

2.2   *High-level roadmap for the programme*

2.3   *Corporate and programme governance requirements*

2.4   *Proposed budget*

2.5   *Programme   management   resources,   roles   and   responsibilities*

2.6   *Risk and opportunity management*

2.7   *Stakeholder management*

A programme charter setting out all the principles as agreed during the framing workshop must be compiled in writing and approved by the programme sponsor. Due to the nature of programmes, the charter may require some changes or updates in terms of the amount of detail as the programme evolves initially. Any potential changes should be tested against the charter and if required to meet the original intent, the charter must be updated, re-approved and communicated. Only once a charter is issued can the owner management team have a kick-off meeting to plan the execution of the shaping process. The content of a programme charter is discussed in detail in Chapter 4 on programme initiation and shaping.

Alignment

Alignment, on the other hand, aims to get the programme team in a position to continue with the work planning in a focused way. Alignment thus involves:

• Informing the team(s) of the requirements as stipulated in the programme charter, and;

• Work-shopping an outline plan of how to achieve the objectives as set out in the charter.

On a programme, the phases (shaping, planning, organising, control, launch and review) as described in this book overlap and are not mutually exclusive linear processes. Programme definition thus evolves over an extended period of time. The alignment process cannot be a singular event and specific alignment interventions need to be planned carefully to firm up the intent throughout all the programme phases. Alignment planning should integrate into the overall programme roadmap and be diligently followed and managed on a continual basis.

In order to achieve alignment, the programme team needs to ensure that they know the:

• **Scope:** what they are supposed to do;

• **Business objectives:** why they are supposed to do it;

• **Programme objectives/operating model:** how they are going to do it;

• **Timeline:** when they are to do it;

• **Roles and responsibilities:** who will be responsible for what;

• **Budget and cash flow:** what monies will be available to the programme, and;

- **Communication plan:** how they will communicate about the programme.

Alignment is a step-wise process that requires interfacing with various key internal stakeholders. Typically, alignment will be required between:

- The sponsor and the programme approval authority;

- The sponsor, the programme management team and the business managers from the various divisions;

- Within the programme management team itself;

- With the different owner support groups;

- With contractors and principal suppliers, and;

- With external stakeholders.

The focal points and levels of interest are different for each of these groups and should be tackled separately by the sponsor and programme management team.

Internal alignment has to be obtained before an attempt is made to gain external alignment. For example, in any holding company with multiple business units each tend to have their own priorities. As part of the internal alignment process, these businesses have to agree on the overall business objectives, as well as how these are translated into project objectives.

Key success factors for alignment

Maintaining alignment will result in a more harmonious working environment with the team remaining focused and all stakeholders staying up to date with the programme objectives, progress and achievements of the programme to date. Paying attention to the key success factors below will support the effectiveness of the alignment effort:

- The right stakeholders are appropriately represented on the programme management team;

- Programme leadership, and especially the sponsor, is mobilised, effective and accountable;

- The priority between cost, schedule and quality drivers is agreed, well defined, communicated and understood by all working on the programme;

- Planning tools and processes are effectively used, for example the framing process;

- The team culture fosters trust, honesty and shared values;

- Communication within the team and with respective business owners is open and effective;

- Effort is put in to clarify roles and responsibilities;

- Team meetings are timeous and effective, and;

- Team work and team building programmes are effective.

Alignment on a programme in context

Achieving and maintaining alignment throughout a programme requires focus and planning in multiple dimensions. Figure 13.2 attempts to explain the dynamic nature of alignment on a programme. Three important issues need to be highlighted, namely:

- Alignment should be revisited throughout the programme lifecycle;

- Alignment within the organisation should be established from the programme approval body, to the sponsor, to managers of the respective impacted business units, to the programme team, to functional teams in support of the programme and finally to contractors (top-to-bottom), and;

• The various projects within the programme should continually be aligned towards the overall programme objectives (cross project).

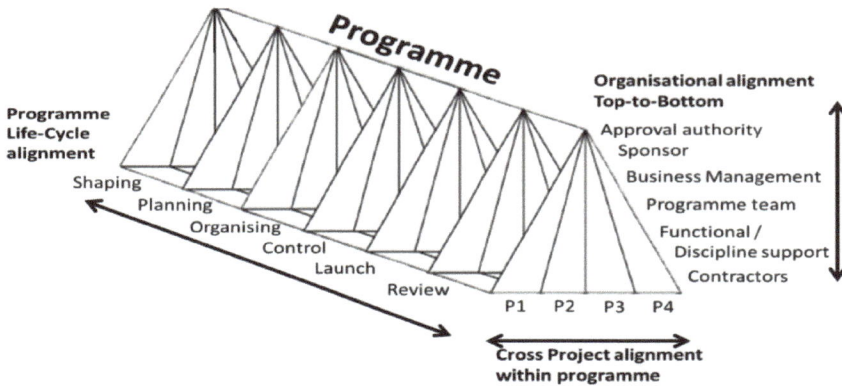

Figure 13.2: **Alignment in context** (adapted from Griffith & Gibson, 1997)

Even though the programme lifecycle phases are overlapping and continuous, it still helps to contextualise certain alignment interventions that mostly fit into a certain phase. The main aim of alignment at programme level and from an owner team perspective, is to make sure everyone remains updated as to where the programme is heading and also of changes to the programme that may come to play. It is also ensuring that the team keeps the long-term business and programme objectives in mind.

Some challenges during the alignment process

A programme is generally prone to some change as shaping develops. Therefore one should attempt to get the best clarity possible from the outset. In the first round, one has to really play devil's advocate to ensure all possible impacts are thoroughly considered. Internal alignment should make provision for discussion of potential questions that external parties may have. It also helps to prepare a list of possible questions and answers.

It is interesting to note that such an alignment workshop initially typically takes at least two weeks of preparation by the chairman and facilitator and at least two to three days' work-shopping the framework agenda with the key decision makers. To further enhance and maintain alignment, informal networking needs to be done by the sponsor and key programme team members on a regular basis, over and above formal framing and alignment sessions.

Achieving alignment during framing may sound like an easy process, but it is not. Each business unit has its own drivers which has a major impact on their priorities. For example, a business unit under financial pressure would want to focus on optimisation and streamlining of its business processes rather than embark on emission reduction programmes.

A challenge during this work is availability of senior management. To get these key players together to agree on the programme objectives can be a daunting task. It requires determined leadership by the sponsor and company executives to drive programme success. A dedicated executive sponsor, who believes in the programme intent, and who has the right level of seniority in the organisation to influence senior management, is essential.

The programme management team should also be giving special attention to alignment when new team members join the team or when requirements change. New members can be brought on board using an on-boarding session. If major changes have been introduced, campaigns should be considered that could include group communication sessions, newsletter articles, bulletins, website notices, bulk texting, road shows and so forth.

Project managers tend to have a very task focused outlook. Therefore it is recommended that a non-technical person with a facilitation and communication background, who understands human interactions, attends key coordination meetings such as steering committees, in order to pro-actively identify issues through the course of a project that may need alignment intervention.

In the next part of this chapter, the alignment approach during programme shaping, programme planning and programme launch will be contextualised.

Alignment during different programme phases

Alignment during programme shaping

During shaping the effort involves aligning the programme team with the business needs of the owner and developing the programme context. The decision makers need to participate during a framing meeting to develop the first issue of the programme charter.

A programme charter is also called a purpose statement, vision or mission and provides a programme with an anchor or organisational focus. A charter can be a powerful tool for focusing the programme on actions and decisions that can have a positive impact on overall success. Participation by all is essential, as taking shortcuts during the shaping phase can cause gaps in the charter due to oversights or inadequate information. This can in turn require changes at a very late stage, leading to cost and schedule overruns.

We now return to the Intego Environmental Compliance Programme to see what the framing process involves for the programme shaping phase. The agenda for the two day framing workshop is shown below as Table 13.1. This is followed with a step-by-step description of the framing process followed by Intego for the initiation and shaping phase of IECP.

Intego Environmental Compliance Programme (IECP)

Agenda for Programme Framing Workshop

The agenda for the two-day IECP framing workshop for the shaping phase of the programme is shown in Table 13.1.

Table 13.1: Agenda for IECP Framing Workshop

Intego Environmental Compliance Programme Framing Workshop Agenda Date:		
Meeting Objective(s): • Develop an aligned view of the business requirements and reasons for embarking on the programme. • Develop an outline of the way forward, resources and requirements.		
Activity & Time	**Session Purpose**	**Discussion Leader**
Day 1: Introduction & Business Overview		
Introduction (09h00 – 09h30)		
• Introduce participants (15 min) • Establish ground rules (10 min)	• Meet one another • Agree standard behaviour and how we will function as a	Facilitator

• Discuss meeting objectives (10 min) • Review and accept agenda (10 min)	group • Ensure agreement on objectives/deliverables • Ensure set agenda will achieve objectives set	
Business overview (09h30 – 17h00)		
• Background/Introduction (30 min)	• Achieve common understanding of the present situation	Sponsor
• Overall vision (60 min)	• Develop an inspiring vision that can be used within the programme team and the business itself that will provide the necessary excitement, inspiration and impetus to drive the programme through to successful completion	Interactive work session by Facilitator
Tea Break (15 min)		
• Business objective(s) (90 min)	• Ensure input and buy-in by all	Sponsor
Lunch (45 min)		
• Programme scope (45min) • Boundary conditions and constraints (15 min) • Major assumptions (15 min) • Potential influences (15 min) • Risks (30 min)	• Obtain input into the initial views and developed a more complete picture of the internal business environment	Intego Holdings Business Development Manager. Facilitator for specific work sessions.

• Exit indicators/Abandonment criteria (15 min) • Impact on key business units (30 min)		
Tea Break (15 min)		
• Key stakeholders (20min) • Community relations and corporate social investment (10 min) • Legislative and governmental SH&E requirements (30 min)	• Obtain input into the initial views and developed a more complete picture of the internal business environment	Interactive work session by Facilitator
Day 2: The Way Forward		
The way forward (09h00 – 17h00)		
• Present level of development of the programme and deliverables completed (30 min)	• Develop a common understanding of the work done to date	Intego Holdings business development manager.
• Programme management principles (30min)	• Agree overall methodology that will be used to manage and govern the programme.	Programme Director
• High-level Roadmap (90 min)	• Establish a first view of what the overall programme schedule could look like • Establish business milestone requirements and must haves	Programme Director & Business Manager
Tea Break (15 min)		

• Programme management resources, roles and responsibilities (45 min)	• Obtain alignment regarding the initial view of the resources required for the programme management (not individual projects)	Programme Director
• Proposed budgets. (45 min)	• Obtain input and alignment of programme study budget and annual running cost • Obtain input and alignment on initial view of total programme capital costs	Programme Director Technical Director
Lunch (45 min)		
• Risks and opportunities (60 min)	• Review existing risks and opportunities • Brainstorm to develop comprehensive list • Prioritise risks and opportunities	Facilitator
• Stakeholder management (60 min)	• Develop stakeholder matrix • Agree stakeholder management strategy	Sponsor.
Closure (17h00 -18h00)		
• Next steps	• Agree actions, responsibilities, timeline for completing Programme Charter	Sponsor

• Check if meeting objectives have been met	• Compare what was achieved with the initial objectives of the framing session	All participants
• Adjourn meeting		

Intego Environmental Compliance Programme (IECP)

Framing the IECP

Step 1: Internal framing session with key stakeholders

A facilitated internal framing workshop between key Intego business stakeholders was conducted. The Sponsor drafted his thoughts on the terms of reference as outlined in Chapter 4. The workshop consisted of the programme sponsor and business executives from Intego Holdings, MiningCo, PowerCo, the programme director and the technical director. The PipelinePro and GridCo JV partners were not involved at this meeting as their direct involvement is small. External parties such as contractors were not included either. Lastly, a technology advisor and a legal advisor were included due to the nature of the programme.

The elements as per the agenda (in Figure 13.1) were discussed and unpacked item by item and consensus reached on each.

Step 2: Finalising the programme charter

The agreements reached during the framing session were then captured as the Programme Charter and Programme Execution Approach. Refer to the final charter as presented in Chapter 4 on initiation and shaping.

Step 3: Alignment with additional key stakeholders

The overall impact of the Intego programme will extend beyond the key decision makers involved in the framing session. For example, the target of reducing SOx, NOx and particulate matter by 10% in 2015, 20% in 2020 and 30% in 2025 is likely to impact every single one of the businesses in the value chain. MiningCo, Pipeline Pro, PowerCo and GridCo, as well as Logistics Pro and UtilityCo may be required to change their business strategies and technology strategies to comply with these targets.

In order to ensure a charter, business strategies and technologies that are aligned to one another, a further workshop was conducted involving all relevant business stakeholders.

The agenda of such a broader alignment session will be determined by the specific relationships and programme needs, but needs to focus on eliciting support from the other owner parties. The most important level of alignment is to agree on the business objectives and the assumptions underlying these objectives.

Step 4: Signing off the charter

The draft charter was updated based on the input from the broader alignment work and approved.

Step 5: Maintaining alignment

Quarterly alignment sessions covering the charter and the assumptions, progress and changes required (if any) were held. This included the senior managers of the owner team and executives from the affected business units.

It is essential to engage all the business decision makers as early as possible to avoid any unnecessary changes at a later stage. Involving the right persons with the right level of authority is crucial.

Proper planning, preparation and facilitation of a framing workshop are crucial to the successful conclusion of the work. See the facilitation guidelines on the OTC Toolkits website: www.otctoolkits.com.

Alignment during planning and organising

Once planning and organising starts it requires involvement of programme management functional personnel like planners, cost estimators and cost controllers, as well as individual project team leads and contractors as appropriate. Planning involves developing detailed work scope descriptions and schedules as outlined in Chapter 5 and organising involves setting up the programme management team (refer to Chapter 6). Personnel from the different business units being affected also start being involved in these activities

To ensure focused effort from all these parties it is essential that the rationale behind the programme as laid out in the programme charter and the approach to be followed as per the programme execution approach be fully understood. Secondly, the activities, roles and responsibilities need to be clarified.

Kick-off meetings are aimed at achieving alignment on the programme objectives and planning the work in terms of owner team deliverables, resources required, and time frames. During the first part, the programme objectives will be introduced and discussed and any uncertainties clarified. The objective now is to clarify understanding of the programme charter and not allow it to be questioned yet again. As the objectives in the charter may be more business focused, it is necessary to translate these into specific 'project' objectives that can be acted on by project management personnel.

The team leaders would review the business objectives to ensure they understand the translated project objectives such that the planning and organising of the various projects can start in a focussed manner.

One of the most important responsibilities of programme management is to facilitate the clarification of roles and responsibilities during this alignment process and kick-off meetings. Especially when you have an integrated programme management team, consisting of client and contractor representatives, the roles and responsibilities need to be very clearly stipulated and more importantly, understood!

Role clarity is critical to the planning and execution of the project. Because projects by definition are unique, the roles of each of the key stakeholders and project leaders are defined at the beginning of the project. Sometimes the roles are delineated in contracts or other documents. Yet, even with written explanations of the roles defined in documents, how these translate into the decision-making processes of the project is often open to interpretation.

The traditional RACI model (Responsible, Accountable, Consult and Inform) for allocating roles and responsibilities is not sufficient in the case of a large programme involving multiple parties. One needs to highlight the person(s) who has the accountability to Approve, but also who is accountable to project manage and coordinate (Lead) an activity. A specific activity can have only one person appointed to lead and coordinate that activity. Because activities are so complex and huge in nature, multiple people may be made responsible to execute tasks that make up the activity. The traditional Inform and Consult remains unchanged. Clear roles and responsibilities avoid duplication and contribute to streamlined operation.

We propose the use of the LACTI model where the acronym has the following meaning:

- **L** = Lead;

- **A** = Approve;

- **C** = Consult;

- **T** = Task, and;

- **I** = Inform.

Returning to the Intego programme, we illustrate the allocation of roles and responsibilities for the programme management team, using the LACTI model.

Intego Environmental Compliance Programme (IECP)

Division of responsibilities for the programme team

A section from the division of work responsibilities for the Intego Programme is shown in Table 13.2 using the LACTI model to illustrate the concept.

Table 13.2: Division of Responsibilities for IECP

Activity	Programme Director	Business Manager	Safety Manager	Programme Manager	Engineering Manager	Controls Manager	Planner	Cost Estimator	Cost Controller
Programme baseline schedule	A	C	I	C	C	L	T	I	I
Project baseline (within programme baseline)	C	C	I	A	C	C	L/T	I	I
Develop & implement programme change management work process	A	C	I	C	C	C	I	I	I
Monitor Engineering Contractor's progress and performance	I	I	I	L/T	I	I	I	I	I
Develop programme cost estimates	A	C	I	C	C	L	I	T	I
Provide programme cost control	I	I		I	I	L	I		T
Provide progress and status reports	I	I		C		L	T	I	I
Maintain programme risk register	L	A	C	T	C	C	C	C	C
Maintain sub-project risk registers	I	I		L/T	I				
Work breakdown structure	A	C	I	C	C	L/T	C	C	C
Develop sub-project cost estimates	C	C		A	C	L		T	

Alignment during launch

Launching is the process whereby sub-projects comprising the programme are sequentially launched or systematically put into operation. The focus during this phase should be to keep stakeholders informed as to the outcomes of the sub-projects and progress towards meeting the overall objectives. Please see Chapter 8 on sequential launching for technical details.

Concluding remarks

Key takeaways from this chapter on framing and alignment are as follows:

- The alignment process consists of a framing section and an alignment section;

- The framing elements should be agreed to and signed off by all the stakeholders that can lay claim to benefits from the programme through an interactive process of engagement;

- Framing is not a one-step process to obtain agreement, approval and sign-off. It will take several sessions and continual effort. Verbal agreement is not sufficient. Physical sign-off signifies commitment;

- The outcome of the framing process is the Programme Charter;

- The purpose of the alignment process is to develop a common understanding of the purpose, agree on the means and methods, and establish trust;

- The components of the alignment process are discussions of the purpose, goals, participant roles, methods of tracking progress and costs, as well as methods of managing change;

- During alignment the Charter is translated into programme and project objectives, and;

- The effects of a lack of trust are delays caused by fact checking or missing information that was not shared because the person's discretion was not trusted to handle sensitive information.

In the next chapter, an overview of the principles of corporate and programme governance is presented.

Chapter 14:
Corporate and Programme Governance

"The time is always right to do right" - Nelson Mandela

Introduction

According to Cadbury (1992), corporate governance is concerned with holding the balance between economic and social goals and between individual and communal goals. He maintains that the governance framework is there to encourage the efficient use of resources and equally to require accountability for the stewardship of those resources, with the aim to align as nearly as possible the interests of individuals, corporations and society. 'Governance framework' is the outer shell of the programme management model, as shown in Figure 14.1

The art of good governance is having as few as possible processes and systems in place to prevent malfeasance, whilst encouraging effective growth and innovation to achieve the strategic objectives of the organisation. In this chapter the focus will be on corporate governance as well as project/programme governance. An attempt will be made to highlight the differences and commonalities. The role of ethics in governance and business sustainability will be discussed, as well as the role of the project sponsor in ensuring proper programme governance. Keep in mind that any project or programme is an inherent part of an organisation's activities and therefore all governance principles equally apply to projects/programmes.

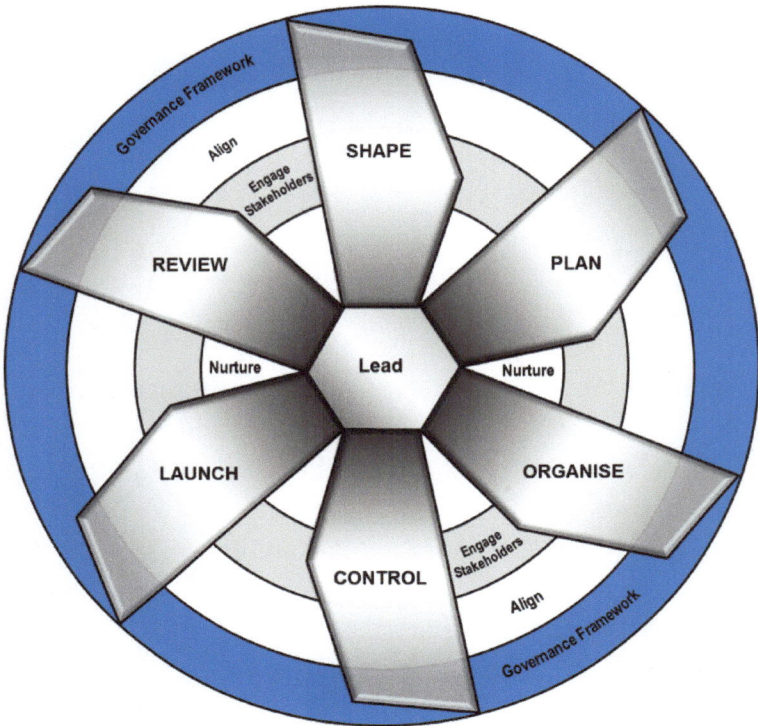

Figure 14.1: Programme management model with 'Governance Framework' highlighted

Corporate Governance

Introduction

This section on corporate governance is based on a series of Insight Articles on our website (Steyn, 2014a, 2014b, 2014c & 2014d)). It is repeated here, with minor improvements, because of the importance of corporate governance and ethical behaviour on the future sustainability of any organisation.

See our website www.ownerteamconsult.com for our free monthly Insight Articles. Insight Articles cover a range of topics regarding project, programme and business management

Definitions of corporate governance

The definition of corporate governance as used in the so-called 'Cadbury report' was given in the introduction. From the King III Report (IoD, 2009b), corporate governance is essentially about effective, responsible leadership and responsible leadership is characterised by the ethical values of responsibility, accountability, fairness and transparency. The King III Code provides a framework for corporate governance for South Africa, based on international best practice (IoD, 2009a).

Bandsuch, Pate and Thies (2008) describe corporate governance as a variety of principles and practices that direct the core processes of a business and state that corporate governance specifically "reflects the formalised values and procedures implemented by the business' leadership in its various operations and interactions with stakeholders".

Archer (2008) maintains that corporate ethics and standards of conduct are matters of governance and the organisation's ethics officer should periodically update the board of directors on the health of the ethics programme, the results of significant investigations and any trends in company conduct that might be emerging. According to Johnson (2005), corporate governance practices are built on the premise that the leaders of companies have an obligation to be fair, transparent, accountable and responsible in their conduct toward shareholders and civil society.

It has been proven that good governance equates to good business. In a study of the relationship between corporate governance compliance and corporate performance, Van der Bauwhede (2009) concludes that greater compliance with international best practices concerning board structure and functioning is significantly and positively correlated with the one-year-ahead return on assets (ROA).

Corporate governance practices are built on the premise that the leaders of companies have an obligation to be fair, transparent, accountable and responsible in their conduct toward shareholders and civil society.

Different approaches to governance

Governance of corporations can be achieved through legislation, through the adherence to a voluntary code of principles and practices or through a combination of the two (IoD, 2009b). The United States of America has chosen the legislated option with legal sanctions for non-compliance. The countries of the Commonwealth and the European Union countries have opted for a combination approach, with an emphasis on the voluntary code of principles and practices. In South Africa, the approach proposed by the King III Report is 'apply or explain', where principles over-ride specific recommended practices. A board of directors could decide to apply a recommendation differently, or follow an alternative practice and still achieve the governance objectives of fairness, accountability, responsibility and transparency. Therefore, explaining why a different set of rules or practices was followed for specific issues, results in compliance. Compliance with governance issues that are legislated is considered essential (IoD, 2009b).

In America, despite the many existing rules, prosecutions and settlements, the response to the corporate scandals of the early 21st century has been primarily rule-based (Michael, 2006). The United States Congress passed the Sarbannes-Oxley Act of 2002 (USA, 2002) and the Securities and Exchange Commission and other agencies have promulgated numerous regulations implementing Sarbannes-Oxley and addressing corporate governance issues. Michael (2006) states that this post-scandal preoccupation with rules and regulations could have been predicted: rules have become substitutes for the 'right' thing when the distinction between obeying the rule and acting ethically becomes blurred and thus 'if something is legal, it's ethical'.

Michael (2006) concludes that rules, statutes and regulations have triumphed over ethics when they become the ceiling rather than the floor for desired ethical conduct. Michaelson (2006) agrees and argues that greater compliance only creates the illusion of ethical progress; "neither goodness without choice nor choice without goodness constitutes ethical progress". Kermis and Kermis (2009) maintain that the Sarbannes-Oxley Act and several other privacy protection laws have not stopped corporate misconduct.

Jennings (2006) warns that one should not automatically equate socially responsible behaviour with good governance and corporate ethics and says that the emphasis on doing good may well be a cover for what goes on internally.

Principles of good governance

The function of governance is, and should be, separate from the function of management although some managers may fulfil both governance and management functions at different times. According to Mosaic Project Services (2014), the core principles of effective governance include the following:

- It is a holistic process focused on the creation of sustainable value by the organisation;

- Authority for some aspects of governance can be delegated to management, however, accountability remains with the governing board;

- Governance and management should be separate; importantly a manager cannot govern his/her own work;

- The governance structure is defined by the governing board and implemented by the organisation's management, and;

- A core aspect of good governance is making the decisions to invest in developing the appropriate management capabilities to ensure that organisational resources are used efficiently and effectively.

Role of Ethics in Governance and Sustainability

What is sustainability?

According to the UN Global Compact-Accenture CEO study (Accenture, 2010), 93% of chief executive officers believe that sustainability is essential to the future success of their companies. Crews (2010) considers sustainability to be more than a fad and states that sustainability involves "creating a permanent shift in the very nature of business". Sustainability currently focuses on an organisation's economic, social and environmental performance. According to Crews there is a tendency to view these three arenas "as contradictory, with competing interests", which results in "a balancing act, in which the emphasis is on trade-offs [...] rather than on a synergistic approach seeking mutual gain".

Dyllick and Hockerts (2002) define corporate sustainability as "meeting the needs of the direct and indirect stakeholders (such as shareholders, employees, clients, pressure groups, communities, etc.), without compromising the ability to meet the needs of future stakeholders as well." They contend that firms have to maintain and grow their economic, social and environmental capital base while actively contributing to sustainability in the political domain. The King III Report (IoD, 2009b) states that sustainability is the primary moral and economic imperative of the 21st century.

Corporate sustainability is meeting the needs of the direct and indirect stakeholders (such as shareholders, employees, clients, pressure groups, communities, etc.), without compromising the ability to meet the needs of future stakeholders as well.

Components of sustainability

The King III Report (IoD, 2009b) states that responsible leaders build sustainable businesses by having regard to the company's economic, social and environmental impact on the community in which it operates. Responsible leaders also consider both the short-term and long-term impact of their personal and institutional decisions on the economy, society and the environment.

According to Beckett and Jonker (2002), sustainability comprises three forms of accountability, namely social-ethical, environmental and economic or financial accountability. Schwartz (2000) believes that a sustainable or 'good' company is defined by three attributes: profitability, meeting the aspirations of stakeholders and having integrity. He maintains that integrity for a company means being well-integrated with one's society and understanding what society's expectations are. Responsible leaders do business ethically rather than merely being satisfied with regulatory compliance, uncritically aligning with the standards of peers or limiting themselves to current social expectations (IoD, 2009b).

These researchers confirm that the components of sustainability are financial viability, corporate social responsibility, and environmental responsibility, all within an ethical framework:

- **Financial Viability:** Financial viability of a company is an obvious requirement; if a business does not generate a positive income, it will not survive for very long. A company needs to create value and make sufficient profit to cover all operational expenses, attract and compensate shareholders and invest in future growth through capital expansion or research and development. In the past, value was narrowly regarded as financial value for stakeholders, but value has come to imply a balanced economic, social and environmental performance (IoD, 2009).

- **Corporate Social Responsibility:** Freeman (2009) proposes an alternate view of the purpose of the organisation, namely "managing-for-stakeholders" or "stakeholder capitalism". Stakeholder capitalism views business as a set of relationships amongst groups who are either affected by, or have an effect on, the business activities, i.e. groups who have a stake in the business and who, by their interaction, create maximum value for all stakeholders. Stakeholder capitalism involves creating as much value as possible for the stakeholders, without resorting to trade-offs. Fombrun and Foss (2004) state that the 'social contract', or bond, linking companies to their stakeholders, is what ensures companies their 'licence to operate'.

- **Environmental responsibility:** Businesses should not expect society and future generations to pay for the social and environmental impacts of its operations. Unethical and irresponsible environmental management can result in stakeholder health and safety concerns, environmental clean-up costs and international challenges such as global warming. The hazards of carbon dioxide emissions and the resulting climate change cannot be neglected or ignored. According to Hauff (2007), those who ignore the effects of climate change are behaving irresponsibly, irrationally and immorally.

The components of corporate sustainability are financial viability, social responsibility and environmental responsibility, all ensconced in an ethical foundation.

Linking ethics, governance and sustainability

Corporate governance is essentially a company's practical expression of ethical standards. As far as legal compliance is concerned, King III states that companies must comply with all applicable laws and remain true to the spirit, intent and purpose of the law.

Sustainability comprises the components of financial viability, social responsibility and environmental responsibility, which are "interconnected in complex ways that should be understood by decision-makers" (IoD, 2009b). Sustainability performance implies that performance in each of these three areas should be integrated and balanced, whilst at the same time meeting all applicable legal requirements. Ethical conduct is essential for every stage of the sustainability framework. The discussion and arguments presented in this section on ethics, governance and sustainability are summarised in Figure 14.2.

Starting from the top of Figure 14.2, the ultimate aim is to ensure a sustainable business. A sustainable business is a function of financial viability, social responsibility and environmental responsibility. In other words, the business has to remain in good standing with all stakeholders and not only the shareholders thereof. It has to generate a profit, whilst being socially and environmentally responsible. Legal compliance is essential, since any business acting outside the law, runs the risk of losing its licence to operate.

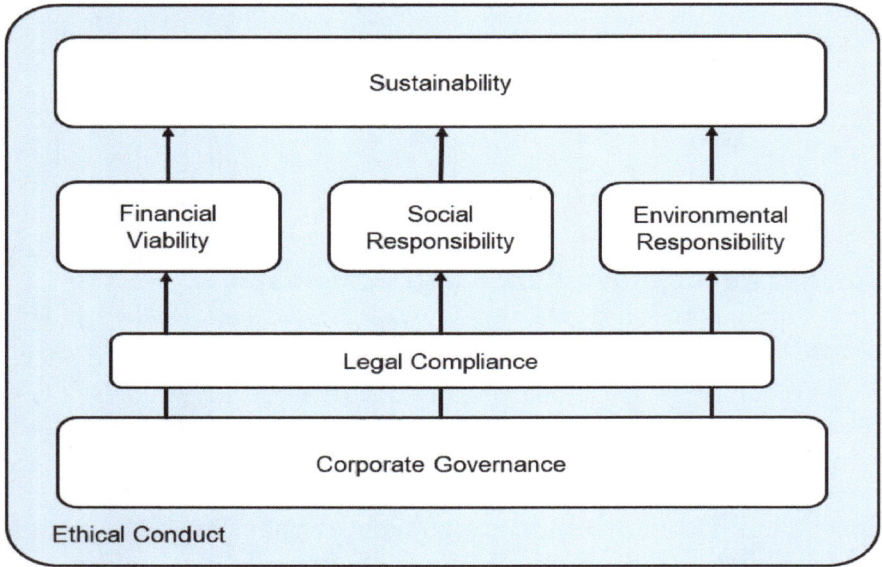

Figure 14.2: Sustainability and ethical conduct

However, the foundation of a sustainable business is effective corporate governance. Corporate governance is essentially the variety of principles and practices that direct the core processes of a business. Each of the elements in Figure 14.2 is driven by ethical decision-making, i.e. decision-making that considers the effect of that decision on all stakeholders, and therefore the elements are all enclosed by the concept of 'ethical conduct'.

The link between ethical conduct and business sustainability is clear. It is essential that businesses with a long-term focus follow an integrated approach to the components of sustainability (financial viability, social- and environmental responsibility). These components of sustainability are not mutually exclusive and need coordinated attention, to the satisfaction of all stakeholders. It is a logical assumption that a similar relationship exists between ethical conduct, programme/project governance and sustainable programmes and projects.

Programme and Project Governance

A subset of corporate governance

Programme and project governance fits within the overall governance of the organisation and is therefore ultimately the responsibility of the Board of Directors. It is a subset of the activities involved in corporate governance as illustrated in Figure 14.3.

The outer circle in Figure 14.3 reflects all the business activities of the organisation. Within this sphere of operation there are governance activities and programme/project management activities. The overlap of these two sets of activities, i.e. the intersection in the Venn diagram, represents governance of programme and project management. Ideally, the programme and project governance principles for an owner organisation will be entrenched in the owner's project/programme management work methodologies.

Programme and project governance refers to a framework that sets out the structure, resources, communication, reporting and monitoring systems to manage a project consistent with an organisation's corporate or strategic vision. In other words, governance of programme and project management defines the framework within which programmes and projects will be conducted.

Programme and project governance encompasses project, programme and portfolio governance and the governance of supporting structures and systems such as the project management office (PMO) and project control boards (PCBs). It focuses on overseeing the management systems that ensure the right projects and programmes are selected by the organisation to achieve its strategic objectives, and that those selected are accomplished as efficiently as possible within the policy framework.

Figure 14.3: Governance of programme and project management (Adapted from APM, 2011)

Programme and project governance refers to a framework that sets out the structure, resources, communication, reporting and monitoring systems to manage a project consistent with an organisation's corporate or strategic vision. It defines the framework within which programmes and projects will be conducted.

The key elements in the governance of programmes and projects include (Mosaic Project Services, 2013):

- Specifying the delegation of authority and responsibility among different participants in the organisation and project management office;

- Defining the rules and procedures for making decisions to ensure that decisions are made at the appropriate level and by competent people;

- Defining the strategic framework needed to select the 'right' programmes and projects to undertake (i.e. portfolio management);

- Encouraging the efficient use of resources;

- Encouraging environmental and social responsibility;

- Monitoring performance of programme and project execution (value assurance);

- Ensuring legal compliance in all programme and project related activities;

- Ensuring proper management support for the organisational change needed to realise the intended benefits, and;

- Requiring accountability at all levels for the stewardship of the resources used.

Programme governance should be fit for purpose

The design of suitable programme governance structures is highly specific and is both organisation and situation dependant. There is no single best practice applicable to all situations. Best practice in terms of governance can help us to consider the factors to be taken into account when putting in place or reviewing governance but what it cannot do is to tell us the right answer for all cases.

A programme extending over five to ten years generally will have a major impact on a company's future and should be set up and managed on the same principles as a major business division of the company. The programme sponsor and programme director should preferably report directly to the chief executive officer or at least an executive director. The programme needs to be set up as a long term venture with offices, infrastructure, personnel and personnel planning, performance agreements, etc.

Programme governance does, however, need to be fit for purpose and some of the considerations include:

- **Programme sponsor:** The sponsor takes responsibility for governance of the programme and is accountable to the company board. The sponsor requires a clear terms of reference and delegated authority (mandate);

- **Company boards:** The board approves the execution of programme through the various stages and issues a mandate to the programme sponsor;

- **Programme steering committees:** Steering committees play an advisory role to the sponsor and require cross-functional membership to reflect the stakeholders in the programme;

- **Project steering committees:** These are required for sub-projects of sufficient scale, complexity and risk. A balance is required when deciding to create subordinate steering committees. These would also imply having sub-sponsors (reporting the programme sponsor) accountable for each steering committee area. The flatter the structure can be kept, the better, and will result in line of sight of the actual activities. Having too many levels of subordinate steering committees results in many meetings and non-alignment between the different projects or areas;

- **Steering committee membership:** Membership must be confined to a small core group of 'experts' that can support the sponsor as well the leadership team of the programme or project. The tendency is to involve everybody who could perhaps make a contribution at some or other time. Large programmes typically attract a lot of attention, with people believing they should be 'seen' on the steering committees. Specific expertise should rather be co-opted on an as needed basis.

- **Enterprise Project Management Office (PMO):** Ensure that the projects and programmes are executed according to the company project related procedures and governance principles;

- **Programme Technical Authority:** Should the programme involve complex technology or highly specialist knowledge, a Technical or Design Authority may be needed to approve scope changes which affect the technical characteristics of the product or capability produced or configuration management changes which impact the target benefits;

- **Programme Team:** The Programme Team is another form of governance depending on the skills of the programme director and the degree of delegated authority from the programme sponsor and board to the programme director. In fact, it is very important that the programme director feels empowered and accountable for the management of risks and issues through his or her team. In a large programme dozens of decisions are being made each week and it is simply impossible for senior management to be involved in them all;

- **Delegation of authority:** In contrast to individual projects, the programme cannot be effectively executed within the 'normal' approval limits. It is advisable to delegate and set up

329

appropriate commercial structures and approval levels to enable smooth turn-around of procurement documents.

Review of programme and project governance

Effective governance of project management ensures that an organisation's project portfolio is aligned to the organisation's objectives, is delivered efficiently and is sustainable. Governance of project management also supports the means by which the board and other major project stakeholders exchange timely, relevant and reliable information.

We have seen that programme and project governance structures and systems should be made fit for purpose. The Association for Project Management (APM, 2011) has developed guidelines for the governance of project management. These guidelines seek to help boards of directors check their organisations against four main components of the governance of project management, namely:

- Portfolio direction and alignment with strategy;

- Project sponsorship;

- Project management capability, and;

- Disclosure and reporting.

These guidelines can be used to good effect in determining the adequacy of the governance structures and systems in place for a specific programme/project. Executing the programme according to a structured stage-gate process allows for proper value assurance and programme gate reviews following each stage.

In the following section we take a closer look at the role of the programme sponsor.

Programme/Project Sponsorship

Role of the sponsor

Effective executive sponsorship is the single most important ingredient required for successful programme and project governance. Without executive sponsorship, most enterprise-wide process initiatives lack a decisive voice capable of resolving process-related conflicts that arise during project implementation.

The project sponsor is the individual who is ultimately answerable to the board for the business and project outcomes of the work (project or programme), in other words the accountable person. This includes holding the ultimate 'yes' or 'no' authority and veto power. Only one accountable person should be assigned to a project or programme and a formal statement should define to whom the sponsor is accountable. In the case of a programme sponsor, accountability for the success (i.e. benefits realisation) of a programme is to the company board.

The project sponsor delegates responsibility for execution to a programme director to deliver the project. A programme director can further delegate responsibility to the project managers of the projects in the programme. The project manager can delegate responsibility for delivery of an element of the project deliverables to a supplier project manager via contract. However, the programme/project sponsor still remains accountable for overall success from inception and until project and business objectives have been met.

Depending on the size and complexity of a programme or project, programme/project sponsors can fulfil the sponsorship role full-time or part-time and may be located at different levels in the organisation. Sponsors are the route through which project managers directly report and from which project managers obtain their formal authority, remit and decisions. Sponsors own the programme/project business case. Competent project sponsorship is of great benefit to even the best project managers.

Effective executive sponsorship is the single most important component required for successful programme and project governance.

Responsibilities of the programme/project sponsor

A programme/project sponsor has three main entities to whom he/she is responsible to, namely to the board, the programme/project manager and programme/project stakeholders.

Responsibility, or rather accountability, to the board includes:

• Ownership of the programme/project business case and charter;

• Ensuring alignment of the programme/project to the company strategy;

• Focusing on the realisation of business benefits;

• Recommending opportunities to reduce cost and optimise benefits, where it makes sense to do so;

• Leadership on company culture and values;

• Ensuring continuity of sponsorship, and;

• Ensuring sound governance in terms of providing feedback, assurances and lessons learnt.

Responsibilities to the programme director or project manager include:

- Translating business objectives into project objectives;

- Clarifying the business priorities and strategy;

- Clarifying the decision making framework;

- Communicating business direction to the project team;

- Enabling provision of programme/project resources;

- Rapid feedback on decisions taken at board level;

- Managing relationships and engendering trust;

- Supporting the programme director/project manager role;

- Promoting ethical work procedures, and;

- Protecting the project team from external influences and removing stumbling blocks.

Responsibilities to the programme/project stakeholders include:

- Governing internal and external stakeholder relations and communications;

- Accountable for client and stakeholder relationships;

- Arbitrating between stakeholders, if required, and;

- Directing governance of suppliers, partners, clients and users.

What does it all mean in practical terms?

We go back to the Intego Environmental Compliance Programme one last time to illustrate what the above means, in practical terms.

Intego Environmental Compliance Programme (IECP)

Programme governance principles and structure

Governance principles

The following governance and ethical principles were set for the IECP:

- **General approach:** The programme will be completed in full compliance to the ethical code of conduct and governance requirements of the Intego group of companies;

- **Management style:** The management style on the project is characterised by honesty, openness and transparency. All decisions and assumptions are recorded and communicated as required;

- **Environment:** The reason for doing the programme is to improve the environmental performance of Intego. However, a construction and operational phase environmental management plan will be completed as part of the environmental impact assessment and these will be followed to the letter. A system will be put in place for recording environmental incidents during the construction phase. The target is zero incidents;

- **Safety:** It is inherently dangerous to implement changes to existing facilities and constructing new facilities in and around operational sites. The overall aim for the programme is to have zero fatalities and a recordable injury rate of less than 0,2. A safety manager will form part of the programme team

334

and every sub-project will have at least one dedicated safety officer;

- **Health:** Executing a programme of this nature over a number of years will result in stressful situations and the accumulation of stress in individuals. An integrated wellness programme will be introduced with annual medical assessments. An attempt will be made to appoint candidates on the programme team with a higher tolerance for stress;

- **Land risk management:** The overall programme lifecycle continues to eventual shut-down and closure of the facilities. Design of new facilities will be done such that the potential for contamination of soil and groundwater is minimised. A requirement for any land potentially impacted by the programme is a thorough baseline evaluation of groundwater quality;

- **Construction manpower:** Construction manpower will preferentially be sourced from the local communities and, if necessary, trained to meet the required skills levels to execute construction work safely. This will grow the skills-base in the local communities and uplift the individuals involved;

- **Affected manpower:** All efforts will be made to minimise the impact on individuals and to accommodate those affected by plant closures. If necessary, they will be retrained for other positions. Management of change will be used to prepare those affected for coming changes, whilst still getting the best performance from affected facilities;

- **Commercial practices:** All commercial practices and procedures will be in line with current Intego practice and withstand any review. Payoffs and bribes will not be tolerated. No single person will have the power to enter into a contract with service providers and suppliers;

- **Programme scope management:** The scope of the programme and that of the sub-projects will be carefully managed to ensure that focus is retained on meeting agreed business objectives.

Governance structure

In terms of the governance structure for the IECP, the following arrangements can be highlighted:

- **Executive sponsor:** The executive sponsor for the IECP is an executive director of Intego Holdings Limited with appropriate project and business management background. The sponsor position is a full-time position. The sponsor will continue to report to the chief executive officer for the duration of the programme;

- **Company board:** The IHL Board is the approval body for the IECP. The sponsor, as an executive director of Intego Holdings, has a seat on the Intego Holdings Board. The sponsor will give feedback directly to the Intego Holdings Board regarding the progress and risks of the IECP;

- **Programme steering committee:** For the IECP, it was agreed that the programme sponsor will be supported by a programme steering committee to be chaired by the sponsor. The programme steering committee is reflected in the top half of Figure 14.4. Membership of the steering committee will deliberately be kept to a minimum. Specific expertise will be co-opted on an as needed basis;

- **Project steering committees:** Two sub-steering committees will be formed, one for the gas power station projects and one for the coal-fired power stations, as shown in Figure 14.4. Each sub-steering committee will essentially be responsible for decision making on a portfolio of projects. Two additional sponsors will be appointed reporting to the programme sponsor. No individual projects will have specific steering committees and all project managers within each area will report to the respective steering committee.

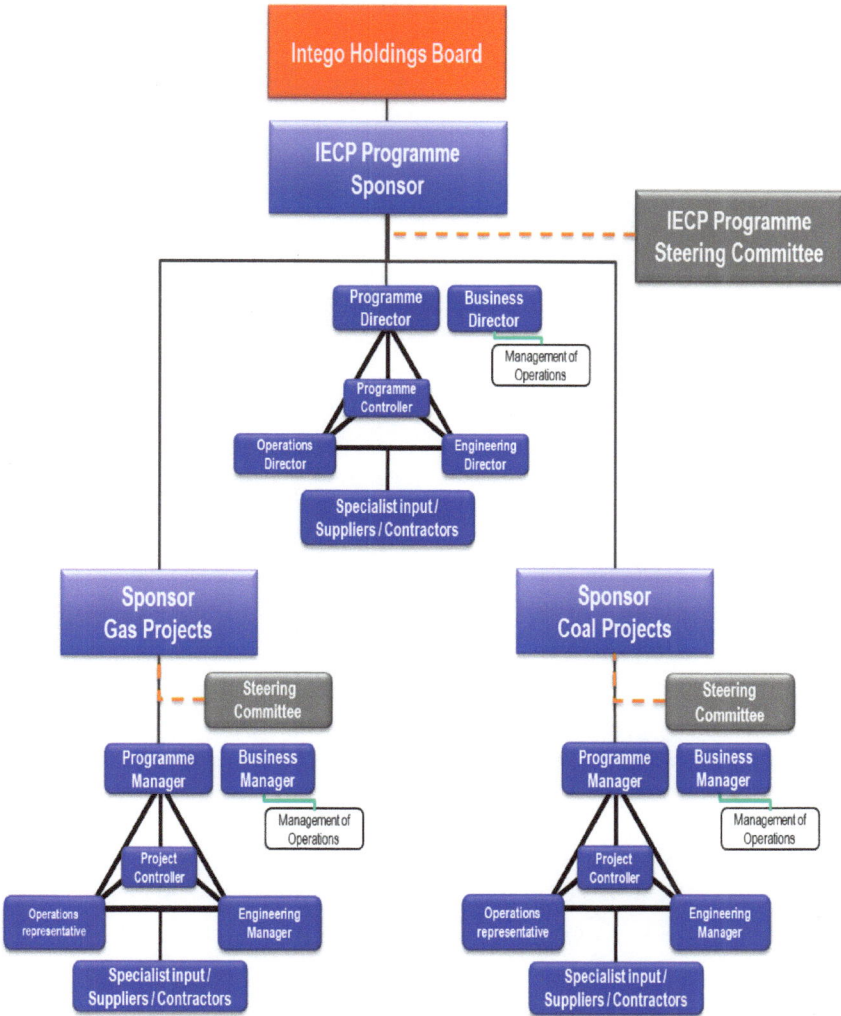

Figure14.4: Sponsorship and steering committees

Although not shown as such in Figure 14.4, both the programme manager for gas projects and the programme manager for coal projects report directly to the programme director. The intent with Figure 14.4 is to show the programme sponsorship structure and not the programme management reporting lines.

The approach of a two-tiered structure necessitates a revision of the overall programme organisation as described in Figure 6.2, such that separate programme managers are appointed for the gas and coal projects respectively. This revised programme organisation, with the two programme managers (the red blocks), is given in Figure 14.5.

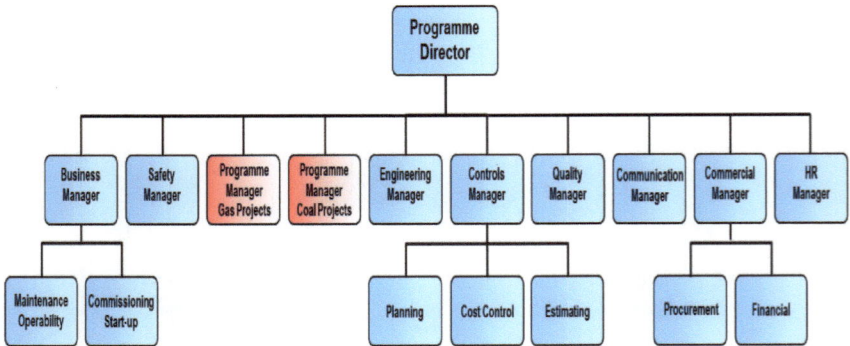

Figure 14.5: Revised programme organisation

- **Programme Technical Authority:** The engineering manager forms part of the revised programme team as illustrated in Figure 14.5. He will be responsible for technology selection, - optimisation and - integration;

- **Delegation of authority:** A representation will be made to the Intego Holdings Board to delegate appropriate approval authority to the programme sponsors and the programme director and programme managers;

- **Project Management Office:** The PMO will be responsible for stage risk reviews, gate readiness reviews and quality assurance deep-dives;

- **Programme Mandate:** The programme mandate is as per the programme charter as discussed in Chapter 4. The programme sponsor agreed to the mandate within which the programme will be executed. The full mandate is shown below for completeness' sake:

-----////-----

Programme Mandate

Intego Environmental Compliance Programme (IECP)

Programme No.:   **12345**

Date: **June 2010**

Prepared and issued by:

_____

Programme Director

Accepted by:

_____

Programme Sponsor

1 Purpose of this document

The purpose of this document is to formalise the delegation of authority and mandate of the Intego Programme Owner's Team as described below in order to minimise the administrative burden on the business.

2 Mandate

The Owner's Team (OT) consisting of the programme director, engineering director and the business representative is mandated to execute the programme within the project baseline as summarised below. The OT needs to highlight deviations from the baseline and obtain approval for significant deviations from the IECP Steering Committee

Approvals are to be dealt with in accordance with the IECP approval and signing authority manual.

2.1 Safety

The overall aim for the programme is to have zero fatalities and a recordable injury case rate of less than 0,2. Recordable case rate (RCR) shall be determined as follows:

$$RCR = \frac{\text{Number of recordable injury cases} \times 200\ 000}{\text{Number of hours worked by programme team members}}$$

A programme safety manager will form part of the programme team. Every sub-project will have a safety manager or at least a dedicated senior safety officer.

If a task cannot be performed safely, it shall be stopped and the approach to the task shall be reconsidered.

2.2   Business Case

2.2.1   Cash flow Management

Cash flow management is an important issue during the execution of the programme.   Should the programme monthly overall cash flow expectation differ significantly from the base case as depicted in Figure 14.6, this will be highlighted on a quarterly forward projection basis.

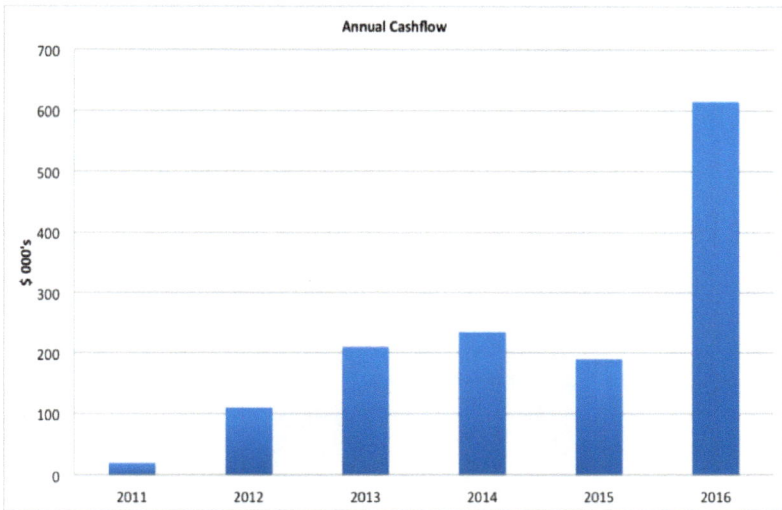

Figure 14.6: Cash flow prediction (* only first 6 years shown)

2.2.2 Risks

Major movements in terms of the previously identified or new risks for the programme will be highlighted to the Steering Committee. Key risks as noted by the board, includes:

- *Interrelationship of sub-projects;*

- *Shutdown schedule risk;*

- *Risks due to brownfield nature of work, and;*

- *Labour force instability.*

2.3 Capital cost

The estimated total capital amount is USD 151 billion for the total programme, which includes a contingency of USD 20 billion. The estimate for phase 1 is USD 940 million, including a contingency of USD 50 million. These capital amounts reflect the upper limit for the programme. As the programme is developed, costs for individual projects are updated and approved. As projects develop, the steering committee must be informed of any requirement to use the overall programme contingency.

2.4 Scope of facility

The high level scope of facilities to be implemented is show in Figure 14.7.

Any additions to or deletions from this high level scope must be approved by the IECP Steering Committee.

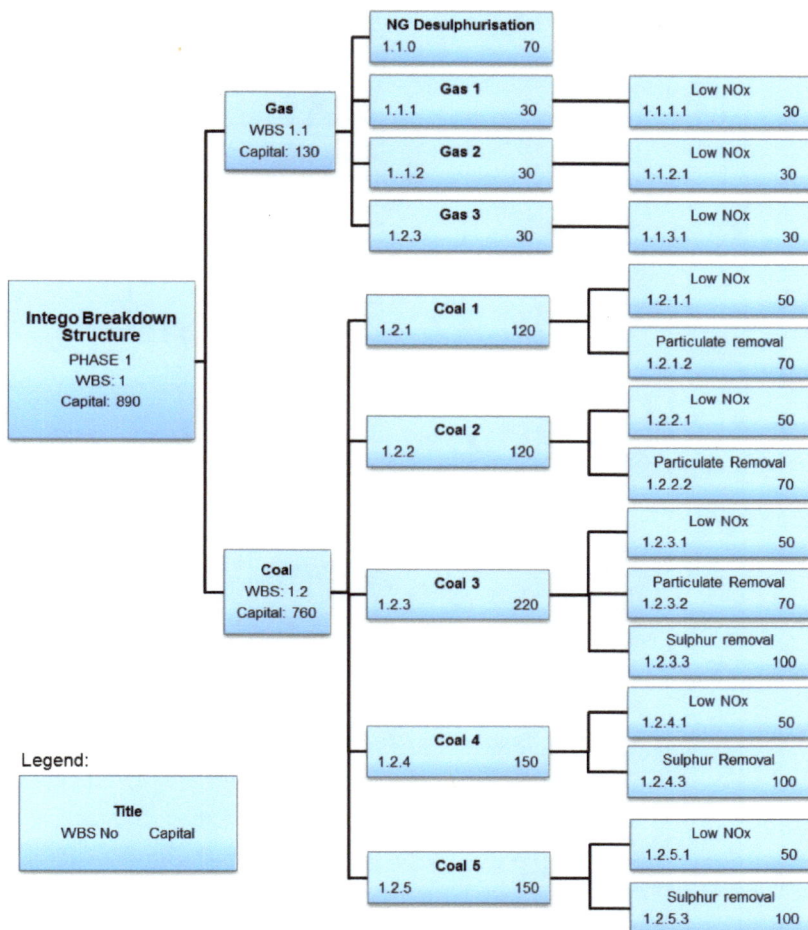

Figure 14.7: High level scope for the IECP

2.5 Baseline schedule

The IECP team shall manage the overall business case in accordance with the baseline schedule as shown in Figure 14.8.

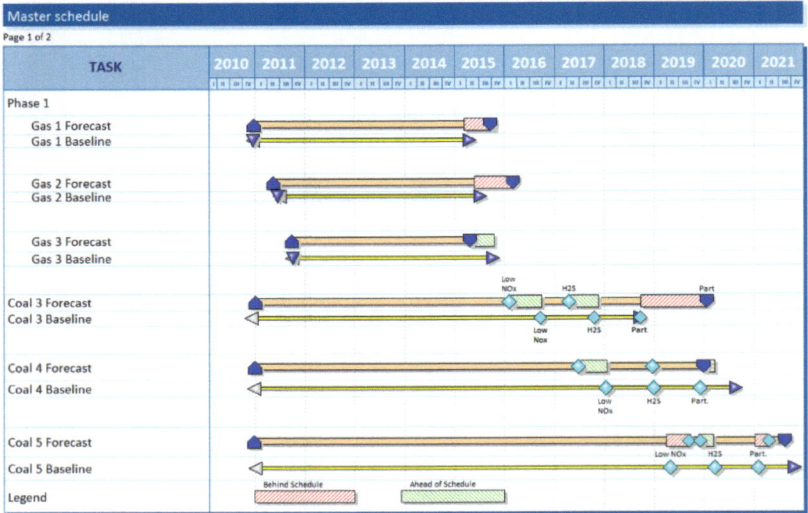

Figure 14.8: Baseline schedule

2.6 Task force principle for human resources

The OT is accountable to ensure that the programme management team is adequately staffed in order to execute the programme. The OT can procure additional resources and demobilise resources as required in order meet their obligations.

Caveat

It is not recommended to try and set a mandate at the outset of a programme. During the initiation and shaping phases, the programme is too fluid and evolving and requires regular interaction between the programme team, the sponsor and the company executive management. Once the overall programme shaping has been completed (refer to Chapter 4 on shaping), a mandate can be set within which the further execution of the programme can be handled.

Concluding remarks

The art of good governance is designing systems that offer sufficient checks and balances to ensure accountability without diminishing the ability of project managers and programme directors to deliver the objectives they have been tasked to accomplish. Governance structures and processes are merely the mechanisms needed to achieve good governance; they do not represent good governance in themselves and should be kept to a practical minimum.

The only question that remains to be answered is the following: If I do everything according to this book, will my programme be a guaranteed success? We attempt an answer in Chapter 15.

This page intentionally left blank

Chapter 15:
Guaranteed Success?

"A project is complete when it starts working for you,
rather than you working for it." -- Scott Allen

Introduction

There seems to be no stopping the tendency for project size and complexity to increase significantly. As stated in Chapter 1, megaprojects are fairly common and, in recent years, even gigaprojects are tackled on occasion. The overriding concern is that projects become so big, complex and interrelated that it is very difficult to grasp the overall picture. The result is that only about 25% of megaprojects are successfully completed (Merrow, 2011). Statistics for gigaprojects are not available, but the assumption is that their success rate will be even lower.

After reading this book and applying the principles as presented, will every programme you tackle be a glorious success? The answer is not necessarily so. There are many extraneous factors and issues beyond the control of the programme team that could influence the eventual outcome. The approach described in this book will however increase the probability of a successful outcome. At the very least, it will make you aware of deviations, how to correct for these and manage stakeholder expectations along the way.

The team of a megaproject or programme has a unique opportunity to manage the programme in such a way that there is visibility on each sub-project rather than just a big amorphous 'programme'.

The F.I.R.E. approach

Ward (2014) describes the F.I.R.E. approach in his book: *F.I.R.E – How Fast, Inexpensive, Restrained, and Elegant methods ignite innovation.* He advocates that projects can be successfully completed using the principles outlined in the book. The question arises as to whether these principles can be applied to programmes as well.

Let us compare his thinking about the F.I.R.E. approach to the principles discussed in this book. We'll show an extract from his book and the comment from the perspective of a programme.

"The F in FIRE stands for fast, which says it is important and good to have a short schedule. It is about defining the project objectives that can satisfy a short timeline, not one we know full well will require twenty years to accomplish."

Once a programme charter has been developed, tackle each sub-project the Dan Ward way. Define the objectives and scope of the sub-project and its boundaries clearly. Get it agreed, plan the project and execute it as efficiently as possible. Handle each sub-project as an as-small-as-possible 'independent' project.

"The I in FIRE for inexpensive says it's important to have a small budget. That may be an unpopular position in an environment in which [big] budgets equal prestige, but in my experience I've found that the ability to deliver meaningful capabilities on a shoestring is actually a widely respected skill, even in cash-rich defence business."

Tackle each sub-project with creativity; understand its business case and use brainpower, not financial power, to develop a solution that will delight the user. This is almost always possible. Break through the perceived barriers and enjoy the result. We concur with Ward when he states "just as fast does not mean hasty, inexpensive does not mean cheap. What we're trying to do is maximize bang for buck. Get the most rumble for our roubles, the most sound for our pound."

"The R of FIRE stands for restrained. It is a preference for self-control, tight budgets, and small teams, for short schedules, short meetings, and short documents."

Make sure the scope of each sub-project is only what is required, no more, no less. Question each feature and functionality that has been requested and determine if it cannot be eliminated by clever redesign. Get the scope approved and carefully manage and control scope, schedule and cost. Have small focused teams for each project and hold them accountable. Do not accept a mediocre result.

"The E in FIRE stands for elegant, in the sense of pleasing, ingenious and simple. Simplicity is an ironically complex topic. It is important and good to have a low level of complexity. A certain degree of complexity is inevitable in any situation. While we may not be able to avoid complexity entirely, we can certainly take steps to minimize it. For simplicity to be elegant and virtuous, it needs to improve the project's quality, performance, or usability."

Many engineers are generally proud to state that the project they are working on is very complex. Does it have to be complex or is it the mind-set of recent day engineers to add more and more complexity to overcome problems, rather than eliminate the cause of a problem? Keep asking yourself how I can make it more intuitive, simpler, easier to use. Keep the objective and scope in mind and compare the design continually with the original intent. Use proven sub systems not novel, first-of-a-kind designs.

In summary, tackle an enormous and 'amorphous' programme a single (FIRE-based) piece at a time. You should have fun, celebrate each achievement, and receive your fair share of reward in the end.

Concluding remarks

There is an old joke about eating an elephant: How does one eat an elephant? The answer, of course, is one bite at a time.

If schedule is not important, you can divide up your elephant in bite-sized portions and feast for many months – hopefully you have a large freezer to keep the meat from spoiling... However, if schedule is important, you will have to obtain help in devouring the elephant. Call upon your friends with healthy appetites to assist you in this gastronomic task. The shorter the time available, the more friends you will require to accomplish the task.

Megaprojects can be as intimidating as the elephant. By following a programme approach and dividing the project up into meaningful bite-sized sub-projects, the task can be accomplished more easily. You will have to draw on the appropriate resources for assistance, because schedule is always an issue. The practical guidelines presented in this book will certainly help you to make a success of your programme.

All that remains is to wish you: Good Luck!

Refer to the Owner Team Consultation website, www.ownerteamconsult.com and our toolkits website www.otctoolkits.com for help on proper execution of programmes and projects, practical tips, templates and advice.

References:
Giving recognition where it is due

"A brave man acknowledges the strength of others."–
Veronica Roth

Accenture., 2010, UN Global Compact-Accenture CEO study: A new era of sustainability. Available from: https://microsite.accenture.com/sustainability/research_and_insights/Pages/A-New-Era-of-Sustainability.aspx (Accessed 28 November 2010).

APM (Association for Project Management), 2006, *Association for Project Management Body of Knowledge (APM-BOK) 5th edition.*, Butler & Tanner, Frome, Somerset.

APM (Association for Project Management), 2011, *Directing Change: a guide to governance of project management 2nd edition.*, Association for Project Management, Princes Risborough, Buckinghamshire.

APM (Association for Project Management), 2012, *Association for Project Management Body of Knowledge (APM-BOK) 6th edition.*, Butler & Tanner, Frome, Somerset.

Archer, D., 2008, Ensuring an ethical organization. *CMA Management,* 82(7):32-36.

Bandsuch, M., Pate, L. & Thies, J., 2008, Rebuilding stakeholder trust in business: an examination of principle-centered leadership and organizational transparency in corporate governance. *Business and Society Review*, 113(1):99-127.

Bartlett, C.A., & Ghoshal, S., 1990, *Matrix Management: Not a Structure, a Frame of Mind,*. Harvard Business Review 68, no. 4 (July–August 1990): 138–145.

Bashir, A., 2007, *Employees' Stress and Its Impact on Their Performance,* First Proceedings of International Conference on Business and Technology, Pages 156-161,Iqra University, Islamabad.

Beckett, R. & Jonker, J., 2002, AccountAbility 1000: a new social standard for building sustainability. *Managerial Accounting Journal,* 17(1/2):36-42.

Brown, J.T., 2008, *The handbook of program management: how to facilitate project success with optimal program management.,* McGraw-Hill, New York.

Buttrick, R., 2010, *The project workout: the ultimate handbook of project and programme management.,* 4th *edition,* Financial Times Prentice-Hall, London.

Cadbury, G.A.H., 1992, *The financial aspects of corporate governance.,* The Committee on the Financial Aspects of Corporate Governance, Gee and Co. Ltd., London.

Campbell, C.A. & Campbell, M., 2013, *The New One-Page Project Manager, Communicate and manage any project with a single sheet of paper, 2nd edition,* John Wiley & Sons, Hoboken, New Jersey.

CII (Construction Industry Institute), 2014, *The Owner's Role in Project Success.,* Construction Industry Institute Publication RS204-1, Austin.

Creagan, E.T., 2012, *Stress can change your personality.* Available from http://www.mayoclinic.org/healthy-living/stress-management/expert-blog/stress-and-personality/bgp-20055977 Accessed on 19 February 2015.

Crews, D.E., 2010, Strategies for implementing sustainability: five leadership challenges. *SAM Advanced Management Journal,* 75(2):15-21.

Deschamps, J-P. & Nayak, P.R., 1995, *Product Juggernauts: how companies mobilize to generate a stream of market winners.* Harvard Business Review Press, Boston.

Dyllick, T. & Hockerts, K., 2002, Beyond the business case for corporate sustainability. *Business Strategy and the Environment,* 11(2):130-142.

Farrelly, F. & Brandsma, J., 1989, *Provocative Therapy 5th ed.* Meta Publications, Capitola, CA.

Forrester, F., 2012, *How your personality affects the way you deal with stress.* Available from: http://www.purelifenutrimedics.com/blog/how-your-personality-effects-the-way-you-deal-with-stress Accessed on 19 February 2015.

Fombrun, C. & Foss, C., 2004, Business ethics: corporate responses to scandal. *Corporate Reputation Review,* 7(3):284-288.

Freeman, R.E., 2009, Managing for stakeholders. In *Ethical Theory and Business, 8th edition.* Edited by Beauchamp, T.L., Bowie, N.E. & Arnold, D.G. New Jersey: Pearson Prentice Hall.

Griffith, A.F. & Gibson, G.E.(jr), 1997, Team alignment during pre-project planning of capital facilities. Research report 113-12, Construction Industries Institute.

Hanford, M.F., 2004, *Program management: different from project management.* PDF file downloaded on 28 February 2015 from IBM developerWorks site: http://atabkamprofessionalservices.vpweb.ca/upload/Program%20manage ment_%20Different%20from%20project%20management.pdf.

Hauff, V., 2007, Brundtland report: a 20 years update. Proceedings of the European Sustainability Conference, held in Berlin, Germany, on 3 to 5 June, 2007.

References

Haughey, D., 2001, *A perspective on programme management: decision support information.*, in ProjectSmart, pdf file downloaded on 7 February 2014 from http://www.projectsmart.co.uk/programme-management.html

Imtiaz, S. & Ahmad, S., 2009. Impact of stress on employee productivity, performance and turnover; an important managerial issue. Pdf file downloaded from http://www.irbrp.com/static/documents/June/2009/38.Subha.pdf on 20 February 2015.

IoD, 2009a, *King code of governance for South Africa 2009.* Sandton: Institute of Directors in Southern Africa.

IoD, 2009b, *King report on governance for South Africa 2009.* Sandton: Institute of Directors in Southern Africa.

Jennings, M.M., 2006, The 7 signs of ethical collapse. *European Business Forum*, 25(2):32-38.

Johnson, K.W., 2005, Integrating applied ethics and social responsibility. Available from: http://www.ethics.org/resource/integrating-applied-ethics-and-social-responsibility (Accessed on 20 October 2010).

Jordan, A., 2009, *Project manager vs. project leader.* Available on http://www.executivebrief.com/project-management/project-manager-vs-project-leader/. Accessed on 4 November 2014.

Kermis, G.F. & Kermis, M.D., 2009, Model for the transition from ethical deficit to a transparent corporate culture: a response to the financial meltdown. *Journal of Academic and Business Ethics,* 2:1-11.

Kerzner, H., 2013, *Project management: a systems approach to planning, scheduling and controlling., 11th edition*, John Wiley & Sons, Inc., Hoboken, New Jersey.

Kiesler, S. & Cummings, J.N., 2002, *What Do We Know about Proximity and Distance in Work Groups? A Legacy of Research,* in

Distributed Work, edited by Hinds, P.J. & Kiesler, S., Cambridge, MA, The MIT Press.

Kloppenborg, T.J., Tesch, D., Manolis, C. & Heitkamp, M., 2006, *An empirical investigation of the sponsor's role in project initiation.,* Project Management Journal 37(3).

Kruse, K., 2013, *What is leadership?* Available on http://www.forbes.com/sites/kevinkruse/2013/04/09/what-is-leadership/. Accessed on 4 November 2014.

LAI (Louis Allen International), 2009, *The Allen management system.* Available on http://www.lai-i.com/almsystem.htm#, Accessed on 25 September 2014.

Leonard-Barton, D. & Swap, W.C., 2005, *When Sparks Fly: Harnessing the power of group creativity.* Harvard Business School Press, Watertown, MA.

Mathis, K., 2013, *Six ways to give proper project leadership.* Available on http://www.projectsmart.co.uk/six-ways-to-give-proper-project-leadership.php. Accessed 4 November 2014.

Merrow, E.W., 2011, *Industrial megaprojects: concepts, strategies, and practices for success.,* John Wiley & Sons, Inc., Hoboken, New Jersey.

Michael, M.L., 2006, Business ethics: the law of rules. *Business Ethics Quarterly,* 16(4):475-504.

Michaelson, C., 2006, Compliance and the illusion of ethical progress. *Journal of Business Ethics,* 66(2-3):241-251.

Mills, H., Reiss, N, & Dombeck, M., Undated, *Types of Stressors (Eustress vs. Distress).* Available from http://www.sevencounties.org/poc/view_doc.php?type=doc&id=15644&cn=117 Accessed on 21 November 2014.

Mindtools, 2014, *Benefits Management: Getting the greatest possible benefit from a project.,* Available on

References

http://www.mindtools.com/pages/article/newPPM_75.htm. Accessed on 27 October 2014.

Mosaic Project Services, 2013, *PPP Governance white paper,* WP1073, Pdf file downloaded from http://www.mosaicprojects.com.au/WhitePapers.html on 14 June 2014.

Mosaic Project Services, 2014, *Governance and Management Systems white paper,* WP1084, Pdf file downloaded from http://www.mosaicprojects.com.au/WhitePapers.html on 14 June 2014.

Murray, A., 2010, *Using PRINCE2 and MSP together – white paper,* The Stationary Office, Norwich, UK.

OGC (Office of Government Commerce), 2009, *Managing Successful Projects with PRINCE2,* The Stationary Office, Norwich, UK.

OGC (Office of Government Commerce), 2010, *The portfolio, programme and project management maturity model (P3M3),* 2nd edition., Axelos Limited, Norwich, UK.

OGC (Office of Government Commerce), 2011, *Managing successful programmes (MSP),* 4th edition., Axelos Limited, Norwich, UK.

Oxford Dictionary, 2014, Oxford University Press, Oxford, UK.

Olsen, G.M. & Olsen, J.S., 2003, *Mitigating the effects of distance on collaborative intellectual work.,* Economics of Innovation and New Technology (12:1).

PMI (Project Management Institute), 2013a, *A guide to the project management body of knowledge (PMBOK® Guide), 5th edition.,* Project Management Institute, Inc., Newtown Square, Pennsylvania.

PMI (Project Management Institute), 2013b, *The standard for program management, 3rd edition.,* Project Management Institute, Inc., Newtown Square, Pennsylvania.

PMI (Project Management Institute), 2013c, *The standard for portfolio management, 3rd edition.*, Project Management Institute, Inc., Newtown Square, Pennsylvania.

Read, H.W., 2004, *The relationship between the owner's project team and the consultant's project team.*, International Platinum Conference 'Platinum Adding
Value', The South African Institute of Mining and Metallurgy.

Schwartz, P., 2000, When good companies do bad things. *Strategy & Leadership*, 28(3):4-11.

Shehu, Z. & Akintoye, A., 2009, *The critical success factors for effective programme management: a pragmatic approach.*, The Built & Human Environment Review, 2(0).

Steyn, J.W., 2014a, Insight article 001: *Business sustainability and ethics part 1: Understanding sustainability.*, PDF file downloaded from http://www.ownerteamconsult.com/blog-1 on 1 December 2014.

Steyn, J.W., 2014b, Insight article 002: *Business sustainability and ethics part 2: Components of sustainability.*, PDF file downloaded from http://www.ownerteamconsult.com/blog-1 on 1 December 2014.

Steyn, J.W., 2014c, Insight article 003: *Business sustainability and ethics part 3: Corporate governance and sustainability.*, PDF file downloaded from http://www.ownerteamconsult.com/blog-1 on 1 December 2014.

Steyn, J.W., 2014d, Insight article 004: *Business sustainability and ethics part 4: Role of ethics in sustainability.*, PDF file downloaded from http://www.ownerteamconsult.com/blog-1 on 1 December 2014.

Steyn, J.W., 2015, Insight article 010: *Projects, programmes and project portfolios.*, PDF file downloaded from http://www.ownerteamconsult.com/blog-1 on 2 February 2015.

Steyn, J.W. & Lourens, D., 2014, Insight article 008: *The Owner Organisation's Project Management Team.*, PDF file downloaded from http://www.ownerteamconsult.com/blog-1 on 1 December 2014.

Treven, H., 2002, Stress results in the work performance. *International Review on Public Studies*, 6(2), 102-118.

USA, 2002, *Sarbanes-Oxley Act of 2002.* H.R. 3763. Washington: Government Printer.

Van der Bauwhede, H., 2009, On the relation between corporate governance compliance and operating performance. *Accounting and Business Research,* 39(5):497-513.

Verma, V.K. & Wideman, R.M., 2002, *Project manager to project leader? and the rocky road between.*, PDF file downloaded from http://www.maxwideman.com/papers/ on 4 November 2014.

Veryard, R., 2001, *In praise of scope creep.*, Requirenautics Quarterly (Newsletter of the Requirements Engineering Specialist Group of the British Computer Society), Issue 24, Pdf file downloaded from http://www.resg.org.uk/articles/ on 23 October 2014.

Ward, D., 2014, *F.I.R.E. – How fast, inexpensive, restrained and elegant methods ignite innovation.*, HarperCollins Publishers, New York.

Weaver, P., 2013, *Project controls – a definition.*, Available on http://mosaicprojects. wordpress.com/2013/11/21/project-controls-a-definition-2/,, Accessed on 16 October 2014.

West, D., 2010, Project sponsorship – an essential guide for those sponsoring projects within their organizations. Gower, Surrey.

-------//////------